Proceedings of the

Gökova Geometry-Topology Conference 2011

Editors

Selman Akbulut
Department of Mathematics, Michigan State University
East Lansing, MI 48824, USA

Denis Auroux
Department of Mathematics, University of California
Berkeley, CA 94720, USA

Turgut Önder
Department of Mathematics, Middle East Technical University
06531 Ankara, Turkey

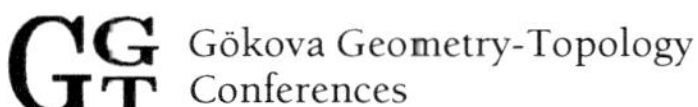
Gökova Geometry-Topology
Conferences

International Press
www.intlpress.com

18th GÖKOVA GEOMETRY-TOPOLOGY CONFERENCE
May 30 – June 4, 2011

Participants of the 18th Gökova Geometry-Topology Conference

PREFACE

This year, in Gökova, we had many interesting talks on a wide variety of subjects from algebraic geometry and mirror symmetry to algebraic topology and handlebodies of 4-manifolds; and also lectures on conformal field theory by the latest fields medalist Stanislav Smirnov. This volume contains interesting articles from this lively eighteenth Gökova Conference. Special thanks goes to Conan Leung for delivering a mini-course on mirror-symmetry and to all the participants for making this conference a very exciting event. We thank TUBITAK and NSF for funding this conference, and thank IP for printing and distributing these proceedings. We also thank Selahi Durusoy for putting these proceedings together, and we finally thank Hotel Yücelen (which is located on the scenic shores of the Gökova Bay) for supporting and hosting this conference.

March 2012

The Editors

CONTENTS

Proceedings of 18th Gökova
Geometry-Topology Conference
pp. 1 – 30

Flat branes on tori and Fourier transforms in the SYZ programme

Kaileung Chan, Conan Naichung Leung and Chit Ma

ABSTRACT. The Strominger-Yau-Zaslow programme says that there should be a theory of fiberwise Fourier transforms along special Lagrangian fibrations which would explain mirror symmetry. This programme has been verified successfully in special cases when quantum corrections do not arise. In this article we will describe a theory of Fourier transforms of flat branes on tori and how it can be applied in the family setting. And in certain special circumstances this explains quantum corrections in SYZ transformations.

1. Mirror symmetry via the SYZ transforms

Roughly speaking, closed string theory on X can be regarded as quantum mechanics on its loop space $\mathcal{L}X$. A path in $\mathcal{L}X$ gives a surface Σ in X. This theory can be interpreted as a 2-dimensional σ-model which studies $Map(\Sigma, X)$. Supersymmetric considerations force X to be a Calabi-Yau manifold. Open string theory concerns paths in X instead, families of paths give surfaces with boundaries. The boundary conditions of these surfaces are specified by geometric objects in X, those satisfying supersymmetry are known as (BPS) branes. In [1], Witten suggests two ways of twisting the σ-model, namely the A-twist and the B-twist. The twisted models are known as A-model and B-model, with corresponding branes known as A-branes and B-branes respectively. The A-model is related to the symplectic geometry of X and the B-model is related to the complex geometry of X. An A-brane in X is a Lagrangian submanifold of X with a flat unitary complex vector bundle on it. Energy minimizing A-branes are special Lagrangian submanifolds. A B-brane in X is a complex submanifold of X with a unitary complex vector bundle (or more generally, a complex of coherent sheaves), such that its connection defines a holomorphic structure on the bundle. Energy minimizing B-branes are Hermitian Yang-Mills bundles.

The mirror symmetry conjecture says that the symplectic geometry (A-model) of a Calabi-Yau manifold X is equivalent to the complex geometry (B-model) of another Calabi-Yau manifold $\check{X}$, and vice versa. Kontsevich [2] formulated the conjecture as an equivalence between the category of A-branes and the category of B-branes, which is known as the homological mirror symmetry (HMS) conjecture. It is also generalized to

Key words and phrases. Mirror Symmetry, SYZ conjecture.

1

manifolds which are not Calabi-Yau.

Strominger-Yau-Zaslow [3] proposed a geometric explanation to mirror symmetry, using physical arguments. Near a large complex structure limit, the quantum effect should be suppressed. To see the relation between a pair of mirror Calabi-Yau manifolds X and $\check{X}$ (of complex dimension n), one can view $\check{X}$ as the moduli space of points in $\check{X}$ and treat it as a family of energy minimizing B-branes in $\check{X}$, it should correspond to a family of energy minimizing A-branes in X. As $\check{X}$ is filled up by points, X is expected to admit a fibration by corresponding A-branes, namely a special Lagrangian torus fibration over a base affine manifold B (of real dimension n). Fix a point $b \in B$, equipping the fiber torus F_b with different flat unitary connections on the trivial line bundle gives a family of A-branes (of real dimension n), which can be parametrized by the dual torus $\check{F}_b$. Therefore, using the fibration structure of X, it is observed that $\check{X}$ is the total space of the dual torus fibration of X over B. Furthermore, one expects that both A- and B-branes should admit tropical limits. Namely they are families of *flat branes* on tori over tropical subvarieties in B. The mirror correspondence should be a fiberwise Fourier-type transform in that case.

The SYZ proposal has been realized in some special cases [4, 5, 6, 7, 8, 9, 10, 11, 12, 13, 14, 15]. In particular, it can be carried out mathematically in certain semi-flat cases, where the Lagrangian fibration for X does not admit singular fibers and the mirror manifold $\check{X}$ is constructed by taking its dual torus fibration. Fiberwise Fourier transform interchanges the symplectic and complex structures between X and $\check{X}$. For A- and B-branes constructed by taking smooth families of flat branes on fiber tori, Fourier transform can be applied to give an explicit correspondence between them. A brief discussion will be given in section 4.

A typical torus fibration admits singular fibers [16]. If we perform the above naive Fourier transform, we only expect to obtain the duality between A- and B-models in their classical limits. To restore the duality, one needs to include the so called quantum corrections in the A-model side.

Quantum corrections for closed string theory are contributed by holomorphic maps from compact Riemann surfaces into X. There is a well developed theory of Gromov-Witten invariants which allows one to "count" the Riemann surfaces contributing to quantum corrections. Quantum corrections for open string theory are related to open Gromov-Witten invariants defined by Fukaya-Oh-Ohta-Ono in [17, 18, 19], which "count" holomorphic disks mapping into X with Lagrangian boundary conditions. Quantum corrections for open string theory are more complicated, both to define and to compute, than their closed counterpart.

The construction of mirror manifold $\check{X}$ by incorporating quantum corrections of X can be carried out explicitly when X is a toric Fano manifold [9] or a non-compact toric Calabi-Yau manifold [8]. In both cases, X admits a "good" torus fibration $\mu : X \to B$ to an integral affine manifold B. The quantum corrections involved are so called genus zero one-pointed open Gromov-Witten invariants, with boundary loops lying in a fiber torus of the fibration. There is a lattice bundle Λ over B parametrizing fiberwise homotopy classes of loops. Generating functions on Λ are defined by "counting" holomorphic disks bounded by a fiberwise loop in Λ. The Fourier transforms of these functions are used to correct the semi-flat complex structure to give a mirror manifold $\check{X}$. In other words, these generating functions appear as "higher Fourier modes" of the corrected mirror complex structure. We give a brief review for these constructions in section 5.

Despite having an explicit construction in the above cases, the reason for using quantum corrections coming from holomorphic disks to correct the mirror complex structure from the semi-flat complex structure is not completely clear. Furthermore, the general procedure for using quantum corrections in the construction of mirror complex structure is not known yet, due to the lack of computational methods for open Gromov-Witten invariants. As we have seen that the corrections appear as "higher Fourier modes" in the above cases, in order to get a better understanding of quantum corrections, it would be important to develop a finite dimensional path space model and establish the Fourier transform in the semi-flat case. Moreover, the Fourier transform defined for flat branes can be used to give a mirror correspondence between A- and B-branes, as a geometric construction for the mirror correspondence. We also give an example to illustrate Fourier transforms between branes incorporating quantum corrections in section 5.

An ongoing project of the authors aims at laying a foundation for SYZ mirror transforms by defining the category of flat branes on tori and constructing a Fourier-type transform. The purpose of this article is to review the classical Fourier transform and provide an outline of the theory of flat branes and their Fourier transform [20]. Explicit examples will be given to highlight some features of the Fourier transform. We will also describe briefly how to apply this construction along Lagrangian torus fibers and give a geometric explanation to certain mirror symmetric phenomena.

2. Fourier transform on tori

In algebraic geometry, we have the Fourier-Mukai transform between an abelian variety X and its dual abelian variety $\check{X} = \mathrm{Pic}_0(X)$: Let $\pi : X \times \check{X} \to X$ and $\check{\pi} : X \times \check{X} \to \check{X}$ be the natural projection maps, $\mathcal{P}$ be the Poincaré line bundle (also called universal bundle) on $X \times \check{X}$. Then the Fourier-Mukai transform functor $\mathcal{FM} : D(X) \to D(\check{X})$ given by

$$\mathcal{FM}(\mathcal{S}) = R\check{\pi}_*(\pi^*\mathcal{S} \overset{L}{\otimes} \mathcal{P})$$

is an equivalence of categories, where $D(X)$, $D(\check{X})$ are the derived categories of coherent sheaves on X and $\check{X}$ respectively. In fact, $\mathcal{FM}$ transforms the Pontrjagin product into the tensor product and vice versa.

Also, $\mathcal{FM}$ induces maps $\mathcal{F}_K$ on K-theory and $\mathcal{F}_H$ on singular cohomology, with the following commutative diagram

$$
\begin{array}{ccc}
K(X) & \xrightarrow{\ \mathcal{F}_K\ } & K(\check{X}) \\[2mm]
{\scriptstyle ch}\Big\downarrow & & \Big\downarrow{\scriptstyle ch} \\[2mm]
H^*_{sing}(X) & \xrightarrow{\ \mathcal{F}_H\ } & H^*_{sing}(\check{X}),
\end{array}
$$

where ch is the Chern character map.

We are seeking an analogue transform of $\mathcal{F}_H$ between a real torus T and its dual torus T^*. Using the deRham isomorphism $H^*_{sing}(T) \cong H^*_{dR}(T)$, it becomes a transform between deRham cohomology groups, which will be the classical Fourier-Mukai transform

$$
\mathcal{F}_{cl} : H^*_{dR}(T) \to H^*_{dR}(T^*).
$$

In this section, we are going to review the definition of $\mathcal{F}_{cl}$ and explore how $\mathcal{F}_{cl}$ can be generalized to an isomorphism in the level of differential forms by incorporating Fourier series. We begin by introducing some notations which will be used throughout the paper.

2.1. Notations

Let V be a real vector space generated by a lattice Λ, with corresponding dual space V^* and dual lattice Λ^*. We use $(\cdot, \cdot)$ as the natural pairing between V and V^*. For a subspace $C_{\mathbb{R}}$ of V generated by a sublattice $C_{\mathbb{Z}}$ of Λ, we have the following exact sequence of vector spaces

$$
0 \to C_{\mathbb{R}} \to V \to V/C_{\mathbb{R}} \to 0,
$$

and the following exact sequence of Abelian groups

$$
0 \to C_{\mathbb{Z}} \to \Lambda \to \Lambda/C_{\mathbb{Z}} \to 0.
$$

Then we have a subtorus $C = C_{\mathbb{R}}/C_{\mathbb{Z}}$ in $T = V/C_{\mathbb{R}}$. Given $a \in V/C_{\mathbb{R}}$ (or $a \in T/C$), we denote by C_a the affine subtorus given by translation of C by a.

On the other hand, we have dual exact sequences

$$
0 \to (V/C_{\mathbb{R}})^* \to V^* \to (C_{\mathbb{R}})^* \to 0,
$$

and

$$
0 \to (\Lambda/C_{\mathbb{Z}})^* \to \Lambda^* \to (C_{\mathbb{Z}})^* \to 0.
$$

We denote $(V/C_{\mathbb{R}})^*$ and $(V/C_{\mathbb{Z}})^*$ by $\check{C}_{\mathbb{R}}$ and $\check{C}_{\mathbb{Z}}$ respectively. Then $\check{C} := \check{C}_{\mathbb{R}}/\check{C}_{\mathbb{Z}}$ defines a subtorus in $T^* = V^*/\Lambda^*$. $\check{C}$ is called the mirror object of C and satisfies $\check{\check{C}} = C$. Similarly, we denote by $\check{C}_{\check{a}} = \check{C} + \check{a}$ the affine subtorus in T^*.

4

Suppose y_j's are coordinates of T (resp. y^j's are dual coordinates of T^*). Then, given $\check{a} = (a^1, \cdots, a^n) \in V^*$ (resp. $a = (a_1, \cdots, a_n) \in V$), we denote by

$$\nabla_{\check{a}} := d - 2\pi i \sum_{j=1}^{n} a^j dy_j \qquad (\text{resp.} \qquad \check{\nabla}_a := d - 2\pi i \sum_{j=1}^{n} a_j dy^j)$$

the connection 1-form determined by $\check{a}$ (resp. a) on T (resp. T^*).

2.2. Fourier transform

From the fact that

$$T^* = H^1(T, \mathbb{R})/H^1(T, \mathbb{Z}) \overset{\exp}{\cong} \mathrm{Hom}_{\mathbb{Z}}(\Lambda, U(1)),$$

we can see that points in T^* parametrize flat unitary connections on the trivial line bundle over T, up to gauge equivalence. The roles of T and T^* can be reversed.

To see this concretely, given a point $\check{a} \in V^*$, we equip the trivial bundle $\mathbb{C} \cdot 1$ with the connection form $\nabla_{\check{a}}$. This determines a map from V^* to space of flat $U(1)$ line bundles on T. As $\nabla_{\check{a}}$ is gauge equivalent to $\nabla_{\check{a}+\check{\lambda}}$ for $\check{\lambda} \in \Lambda^*$, the map descends to T^* as an isomorphism.

We can describe the isomorphism using the Poincaré line bundle $\mathcal{P}$. We define the universal connection by the connection 1-form

$$\nabla_{\mathcal{P}} := d + \pi i \sum_{j=1}^{n} (y_j dy^j - y^j dy_j)$$

on the trivial complex line bundle $\tilde{\mathcal{P}}$ over $V \times V^*$, the universal cover of $T \times T^*$. The natural action of $\Lambda \times \Lambda^*$ on $V \times V^*$ can be lifted to an action on $\tilde{\mathcal{P}}$ given by

$$(\lambda, \check{\lambda}) \cdot (y, \check{y}, t) = (y + \lambda, \check{y} + \check{\lambda}, e^{\pi i [(\check{\lambda}, y) - (\check{y}, \lambda)]} t).$$

Taking the quotient of $\tilde{\mathcal{P}}$ by $\Lambda \times \Lambda^*$, we get the Poincaré line bundle $\mathcal{P}$ over $T \times T^*$ which is a nontrivial bundle. The connection $\nabla_{\mathcal{P}}$ descends to $T \times T^*$ as a unitary connection.

We have the universal property that $\nabla_{\mathcal{P}}|_{T \times \{\check{a}\}}$ and $\nabla_{\mathcal{P}}|_{\{a\} \times T^*}$ can be written as $\nabla_{\check{a}}$ and $\check{\nabla}_a^*$ respectively. This provides an isomorphism from T^* to the moduli space of flat unitary line bundles on T, by sending $\check{y}$ to $\mathcal{P}|_{T \times \{\check{y}\}}$.

Also, the curvature of $\nabla_{\mathcal{P}}$ is given by

$$F_{\mathcal{P}} = 2\pi i \sum_{j=1}^{n} dy_j \wedge dy^j.$$

We use this together with the projection maps π and $\check{\pi}$ from $T \times T^*$ to T and T^* respectively to define the following transform for differential forms.

Definition 2.1. The classical Fourier-Mukai transform $\mathcal{F}_{cl} : \Omega^k(T) \to \Omega^{n-k}(T^*)$ is defined by

$$
\begin{aligned}
\mathcal{F}_{cl}(\alpha) &= (-1)^{\frac{n(n-1)}{2}} \check{\pi}_*(\pi^*\alpha \wedge e^{\frac{i}{2\pi}F_{\mathcal{P}}}) \\
&= (-1)^{\frac{n(n-1)}{2}} \int_T \pi^*\alpha \wedge e^{\frac{i}{2\pi}F_{\mathcal{P}}}.
\end{aligned}
$$

For example $\mathcal{F}_{cl}(1) = dy^1 \wedge dy^2 \wedge \cdots \wedge dy^n$.

We may also define the inverse Fourier transform $\mathcal{F}_{cl}^{-1} : \Omega^*(T^*) \to \Omega^*(T)$ by

$$
\begin{aligned}
\mathcal{F}_{cl}^{-1}(\check{\alpha}) &= (-1)^{\frac{n(n-1)}{2}} \pi_*(\check{\pi}^*\check{\alpha} \wedge e^{-\frac{i}{2\pi}F_{\mathcal{P}}}) \\
&= (-1)^{\frac{n(n-1)}{2}} \int_{T^*} \check{\pi}^*\check{\alpha} \wedge e^{-\frac{i}{2\pi}F_{\mathcal{P}}}.
\end{aligned}
$$

Even though $\mathcal{F}_{cl}^{-1}$ is called the inverse transform, $\mathcal{F}_{cl}$ is not an isomorphism and $\mathcal{F}_{cl}^{-1} \circ \mathcal{F}_{cl} : \Omega^*(T) \to \Omega^*(T)$ is just the harmonic projection. Nevertheless, $\mathcal{F}_{cl}$ descends to an isomorphism on cohomologies

$$
\mathcal{F}_{cl} : H_{dR}^*(T, \mathbb{R}) \to H_{dR}^*(T^*, \mathbb{R}).
$$

Under the deRham isomorphism, $\mathcal{F}_{cl}$ is consistent with the earlier discussion on duality of subtori: We first note that $H_*^{sing}(T)$ is generated by homology classes $[C]$'s of subtori C's in T and $H_{sing}^*(T)$ is generated by their Poincaré duals $\mathrm{PD}[C]$'s. Via the deRham isomorphism, $\mathcal{F}_{cl}(\mathrm{PD}[C]) = \mathrm{PD}[\check{C}]$. Furthermore $\mathcal{F}_{cl}(\mathrm{PD}[C_1 \cap C_2]) = \mathrm{PD}[\check{C}_1 + \check{C}_2]$.

The classical Fourier transform $\mathcal{F}_{cl}$ works nicely in the study of mirror symmetry without corrections, see [11] and [13]. However, we have to extend it by incorporating Fourier series in order to obtain an isomorphism on the level of forms which corresponds to quantum corrections in mirror symmetry.

2.3. Fourier series

In the classical Fourier series setting, a function $f : \mathbb{Z} \to \mathbb{C}$ and a function $\check{f} : S^1 = \mathbb{R}/\mathbb{Z} \to \mathbb{C}$ can be interchanged by

$$
f(\lambda) = \int_{S^1} \check{f}(y)e^{2\pi i\lambda y}dy
$$

and

$$
\check{f}(y) = \sum_{\lambda \in \mathbb{Z}} f(\lambda)e^{-2\pi i\lambda y}.
$$

This can be easily generalized to higher dimensional cases for tori and lattices. Roughly speaking, instead of considering functions on T and T^*, we consider functions on $T \times \Lambda$

and $T^* \times \Lambda^*$. The idea of defining a transform between them is transforming functions on T (resp. Λ) to functions on Λ^* (resp. T^*). Let π and $\check{\pi}$ be the projection maps from $T \times \Lambda \times T^* \times \Lambda^*$ to $T \times \Lambda$ and $T^* \times \Lambda^*$ respectively. Then we have the following definition by combining the Fourier series and our earlier Fourier-Mukai transform.

Definition 2.2. Let $(y, \lambda) \in T \times \Lambda$ and $(\check{y}, \check{\lambda}) \in T^* \times \Lambda^*$. The Fourier transform $\mathcal{F} : \Omega^k(T \times \Lambda) \to \Omega^{n-k}(T^* \times \Lambda^*)$ is defined by

$$
\begin{aligned}
\mathcal{F}(\alpha) \;&=\; (-1)^{\frac{n(n-1)}{2}} \check{\pi}_* \left(\pi^* \alpha \wedge e^{\left(\frac{i}{2\pi} F_{\mathcal{P}} + 2\pi i(y, \check{\lambda}) - 2\pi i(\lambda, \check{y}) \right)} \right) \\
&=\; (-1)^{\frac{n(n-1)}{2}} \int_T \left(\sum_{\lambda \in \Lambda} (\pi^* \alpha) e^{-2\pi i(\lambda, \check{y})} \right) e^{2\pi i(y, \check{\lambda})} \wedge e^{\frac{i}{2\pi} F_{\mathcal{P}}},
\end{aligned}
$$

and the inverse transform $\mathcal{F}^{-1} : \Omega^k(T^* \times \Lambda^*) \to \Omega^{n-k}(T \times \Lambda)$ is defined by

$$
\begin{aligned}
\mathcal{F}^{-1}(\check{\alpha}) \;&=\; (-1)^{\frac{n(n-1)}{2}} \pi_* \left(\check{\pi}^* \check{\alpha} \wedge e^{-\left(\frac{i}{2\pi} F_{\mathcal{P}} + 2\pi i(y, \check{\lambda}) - 2\pi i(\lambda, \check{y}) \right)} \right) \\
&=\; (-1)^{\frac{n(n-1)}{2}} \int_{T^*} \left(\sum_{\check{\lambda} \in \Lambda^*} (\check{\pi}^* \check{\alpha}) e^{-2\pi i(\check{\lambda}, y)} \right) e^{2\pi i(\lambda, \check{y})} \wedge e^{-\frac{i}{2\pi} F_{\mathcal{P}}}.
\end{aligned}
$$

Example 2.3. If $f(y, \lambda) = f(\lambda)$, then

$$
\mathcal{F}(f(y, \lambda)) = \mathcal{F}(f(\lambda)) = \begin{cases} \check{f}(\check{y}) dy^1 \wedge dy^2 \wedge \cdots \wedge dy^n & \text{if } \check{\lambda} = 0, \\ 0 & \text{if } \check{\lambda} \neq 0, \end{cases}
$$

where $\check{f}(\check{y})$ is the ordinary Fourier transform of $f(\lambda)$. Conversely, if

$$
f(y, \lambda) = \begin{cases} f(y) & \text{if } \lambda = 0, \\ 0 & \text{if } \lambda \neq 0, \end{cases}
$$

then

$$
\mathcal{F}(f(y, \lambda)) = \check{f}(\check{\lambda}) dy^1 \wedge dy^2 \wedge \cdots \wedge dy^n,
$$

where $\check{f}(\check{\lambda})$ is the ordinary Fourier series of $f(y)$. In general, suppose $I = \{i_1, \cdots, i_k\}$ is a subset of $\{1, \cdots, n\}$ with $i_1 < \cdots < i_k$, $\bar{I} = \{i_{k+1}, \cdots, i_n\} = \{1, \cdots, n\} - I$ with $\{i_1, \cdots, i_k, i_{k+1}, \cdots, i_n\}$ an even permutation of $\{1, \cdots, n\}$, we have

$$
\begin{aligned}
&\mathcal{F}(f(y, \lambda) dy_I) \\
=\; &(-1)^{\frac{n(n-1)}{2}} \int_T \left(\sum_{\lambda \in \Lambda} (f(y, \lambda) dy_I) e^{-2\pi i(\lambda, \check{y})} \right) e^{2\pi i(y, \check{\lambda})} \wedge e^{\frac{i}{2\pi} F_{\mathcal{P}}} \\
=\; &(-1)^{\frac{n(n-1)}{2} + \frac{|\bar{I}|(|\bar{I}|+1)}{2}} \int_T \left(\sum_{\lambda \in \Lambda} f(y, \lambda) e^{-2\pi i(\lambda, \check{y})} \right) e^{2\pi i(y, \check{\lambda})} dy_I \wedge dy^{\bar{I}} \wedge dy_{\bar{I}} \\
=\; &(-1)^{\frac{n(n-1)}{2} + \frac{|\bar{I}|(|\bar{I}|+1)}{2} + |I||\bar{I}|} \int_T \left(\sum_{\lambda \in \Lambda} f(y, \lambda) e^{-2\pi i(\lambda, \check{y})} \right) e^{2\pi i(y, \check{\lambda})} dy^{\bar{I}} \wedge dy_I \wedge dy_{\bar{I}} \\
=\; &(-1)^{\frac{n(n-1)}{2} + \frac{|\bar{I}|(|\bar{I}|-1)}{2} + |I||\bar{I}|} \check{f}(\check{y}, \check{\lambda}) dy^{\bar{I}}
\end{aligned}
$$

where

$$\check{f}(\check{y}, \check{\lambda}) = \int_T \left(\sum_{\lambda \in \Lambda} f(y, \lambda) e^{-2\pi i (\lambda, \check{y})} \right) e^{2\pi i (y, \check{\lambda})},$$

and $dy^{\bar{I}} = dy^{i_{k+1}} \wedge \cdots \wedge dy^{i_n}$, $dy_I = dy_{i_1} \wedge \cdots \wedge dy_{i_k}$.

As in the classical case, given a function $f : \mathbb{Z} \to \mathbb{C}$, $\check{f}(\check{y}) = \sum_{\lambda \in \mathbb{Z}} f(\lambda) e^{-2\pi i \lambda \check{y}}$ may not converge. Therefore, in our case, we have to restrict to those $f(y, \lambda) dy_I \in \Omega^*(T \times \Lambda)$ such that for a fixed $\lambda \in \Lambda$, $f(y, \lambda) = f(y)$ is a L^2-function and for a fixed $y \in T$, $f(y, \lambda) = f(\lambda)$ is a rapid-decay function. We define $\Omega^*_{(2)}(T \times \Lambda)$ to be the subspace of $\Omega^*(T \times \Lambda)$ consisting of all forms with the above property. Then, we have:

Proposition 2.4. $\mathcal{F} : \Omega^*_{(2)}(T \times \Lambda) \to \Omega^*_{(2)}(T^* \times \Lambda^*)$ *is an isomorphism and* $\mathcal{F}^{-1} \circ \mathcal{F}$ *is the identity map on* $\Omega^*_{(2)}(T \times \Lambda)$.

Remark 2.5. As our motivations come from string theory which is roughly quantum mechanics on loop spaces, it is worth noticing that $T \times \Lambda$ is simply the space of all geodesic loops in T. When we generalize this to the family of tori T in a Lagrangian fibration in a symplectic manifold X, we should consider the space of fiberwise geodesic loops in X. In [9], K.W. Chan and the second author used this method in toric setting to obtain the mirror Landau-Ginzburg model predicted by Kontsevich and Hori-Vafa [21].

3. Flat branes in tori

As mentioned in section 1, an A-brane is a pair (L, E) where L is a special Lagrangian submanifold and E is a flat unitary vector bundle E on L, a B-brane is a pair $(\check{L}, \check{E})$ where $\check{L}$ is a complex submanifold and $\check{E}$ is unitary complex vector bundle such that its connection defines a holomorphic structure on $\check{E}$. In a large structure limit, mirror manifolds are expected to admit dual Lagrangian (possibly singular) tori fibrations over a singular affine manifold B [13]. Furthermore, one expects that both A- and B-branes should admit tropical limits. Namely they are families of flat branes on tori over tropical subvarieties in B. According to SYZ, the mirror correspondence is predicted to be a fiberwise Fourier-type transform on these flat branes on dual tori, incorporating with quantum corrections. This would give a geometric understanding for the mirror correspondence. However, the mathematical description for this correspondence is not completely clear yet, much more work is needed to clarify how singular the families of flat branes can be and how to include quantum corrections.

In order to understand the correspondence better, we study flat branes in a single flat torus, which should be thought of a smooth fiber of the Lagrangian fibration. We are going to discuss how to obtain a bijection between the set of flat branes in a torus and its dual via Fourier transform. Furthermore, we will define a complex using the space of mininal geodesic paths from a flat brane to another to study the quantum intersection between two flat branes.

We start by giving the definition of flat branes. Recall that, with a subtorus $C = C_{\mathbb{R}}/C_{\mathbb{Z}}$ in T and $a \in V/C_{\mathbb{R}}$ (or $a \in T/C$), we denote by C_a the affine translation of C by a.

Definition 3.1. A flat brane $\mathcal{B}$ in a torus T is a pair (C_a, E), where C_a is an affine subtorus in T and E is a flat unitary complex vector bundle over C_a.

Remark 3.2. E splits orthogonally into a direct sum of flat unitary line bundles.

Suppose $\mathcal{B}_i = (C_{i,a_i}, E_i)$, $i = 1, 2$, are two flat branes in a torus T, we will define a complex to study the quantum intersection of two branes. In general, we are supposed to study the path space from C_{1,a_1} to C_{2,a_2}. When the torus T is a fiber of a Lagrangian fibration at a large complex structure limit, the intersection theory should localize to the subspace consisting of minimal geodesics, which is a finite dimensional manifold.

Definition 3.3. For two affine subtori C_{1,a_1} and C_{2,a_2}, an instanton from C_{1,a_1} to C_{2,a_2} is a connected minimal geodesic segment $\gamma : [0,1] \to T$ such that $\gamma(0) \in C_{1,a_1}$ and $\gamma(1) \in C_{2,a_2}$.

Passing to the universal cover, an affine subtorus C is lifted to infinitely many affine linear subspaces. An instanton from C_{1,a_1} to C_{2,a_2}, upon lifting to the universal cover V, is a line segment joining an affine subspace covering C_{1,a_1} and to an affine subspace covering C_{2,a_2}, which is orthogonal to both subspaces. Two points in distinct flat branes are said to be intersecting each other in the quantum sense if they are joined by an instanton. The space of instantons is denoted by $M(C_{1,a_1}, C_{2,a_2})$, or simply $M_{1,2}$ if there is no confusion. There are two natural evaluation maps

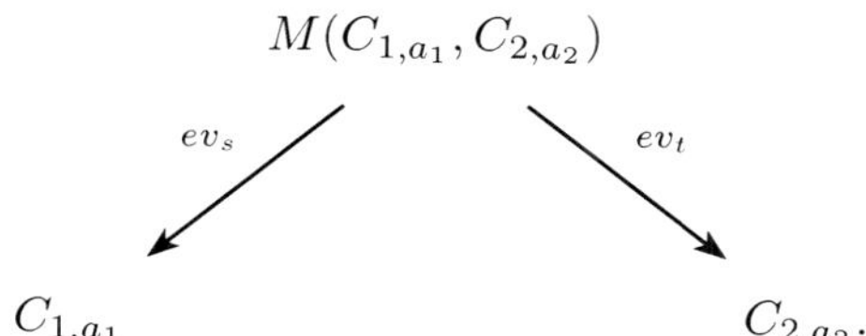

Remark 3.4. In the case that both C_{i,a_i}'s are affine translations of the same torus C, we have a map

$$M(C_{a_1}, C_{a_2}) \xrightarrow{\sim} C_{a_1} \times \Lambda/C_{\mathbb{Z}}$$

given by

$$\gamma \mapsto (\gamma(0), \gamma' + \tilde{a}_1 - \tilde{a}_2),$$

where $\tilde{a}_i$'s are some liftings of a_i's. This gives an explicit parametrization for the space of instantons.

The quantum intersection complex can be defined using the space $M_{1,2}$ in the following manner.

First, we can form the homomorphism bundle over $M_{1,2}$ from $ev_s^* E_1$ and $ev_t^* E_2$, denoted by $\mathrm{Hom}_{qu}(E_1, E_2) = \mathrm{Hom}(ev_s^* E_1, ev_t^* E_2)$. There is a natural flat unitary connection given by $\nabla_{1,2} = ev_s^*(\nabla_1^*) \otimes ev_t^*(\nabla_2)$. This gives a differential complex

$$(\Omega^*(M_{1,2}, \mathrm{Hom}_{qu}(E_1, E_2)), \nabla_{1,2}).$$

Second, we define a bundle $\mathcal{N}_{1\vee 2}$ over the space $M_{1,2}$. Denoting by $P_\gamma : T_{\gamma(0)}T \to T_{\gamma(1)}T$ the parallel transport along the path $\gamma \in M_{1,2}$, we define

$$\mathcal{N}_{1\vee 2}|_\gamma = T_{\gamma(0)}T / (T_{\gamma(0)}C_{1,a_1} + P_\gamma^{-1}T_{\gamma(1)}C_{2,a_2}).$$

The metric on T induces a connection on $\mathcal{N}_{1\vee 2}$ which is flat.

Indeed, if we identify tangent space at every point of T as V, the tangent space of C_{i,a_i} is naturally identified with $C_{i,\mathbb{R}}$. Then we have $\mathcal{N}_{1\vee 2} \simeq V/(C_{1,\mathbb{R}} + C_{2,\mathbb{R}})$ at every path γ. Thus $\mathcal{N}_{1\vee 2}$ is a trivial bundle with trivial connection under this identification.

$s(\gamma) = 2\pi i \gamma' \in \mathcal{N}_{1\vee 2}$ defines a flat section of the trivial bundle $\wedge^1 \mathcal{N}_{1\vee 2}$ over $M_{1,2}$. Therefore, we can define

$$\delta : \Gamma(M_{1,2}, \wedge^q \mathcal{N}_{1\vee 2}) \to \Gamma(M_{1,2}, \wedge^{q+1}\mathcal{N}_{1\vee 2})$$

by $\delta(\varphi) := s \wedge \varphi$. We can verify that $\delta^2 = 0$.

Remark 3.5. There is an equivalent complex defined using the dual bundle $\mathcal{N}_{1\vee 2}^*$, with the differential

$$\check{\delta} : \Gamma(M_{1,2}, \wedge^q \mathcal{N}_{1\vee 2}^*) \to \Gamma(M_{1,2}, \wedge^{q-1}\mathcal{N}_{1\vee 2}^*)$$

given by $\check{\delta}(\varphi) := \iota_s(\varphi)$. The two complexes are identified by an isomorphism $\mathcal{N}_{1\vee 2}^* \simeq \mathcal{N}_{1\vee 2}$ using metric and then the Hodge star operator, up to a factor of $(-1)^{q-1}$ on sections of $\wedge^q \mathcal{N}_{1\vee 2}^*$.

Remark 3.6. The reason for introducing the complex $\wedge^* \mathcal{N}_{1\vee 2}$ will become clear when we will consider families of flat branes. When we consider A-branes, $\mathcal{N}_{1\vee 2}$ is identified with a subbundle of cotangent bundles of two A-branes, using symplectic form. In this case, all non constant paths will contribute and result in a quantum theory. When we consider B-branes, the complex $\wedge^* \mathcal{N}_{1\vee 2}$ is exact off the classical intersection of two B-branes. This gives a classical intersection theory.

We let $\Omega^{p,q}(\mathcal{B}_1, \mathcal{B}_2) := \Omega^p(M_{1,2}, \mathrm{Hom}_{qu}(E_1, E_2) \otimes \wedge^q \mathcal{N}_{1\vee 2})$. Then δ and $\nabla_{1,2}$ induce two commuting differentials

$$\delta : \Omega^{p,q}(\mathcal{B}_1, \mathcal{B}_2) \to \Omega^{p,q+1}(\mathcal{B}_1, \mathcal{B}_2) \text{ and } \nabla_{1,2} : \Omega^{p,q}(\mathcal{B}_1, \mathcal{B}_2) \to \Omega^{p+1,q}(\mathcal{B}_1, \mathcal{B}_2).$$

Therefore, $\Omega^{*,*}(\mathcal{B}_1, \mathcal{B}_2)$ is a double complex. Furthermore, we define the total complex $\Omega^n(\mathcal{B}_1, \mathcal{B}_2) := \oplus_{p+q=n}\Omega^{p,q}(\mathcal{B}_1, \mathcal{B}_2)$ and $D : \Omega^n(\mathcal{B}_1, \mathcal{B}_2) \to \Omega^{n+1}(\mathcal{B}_1, \mathcal{B}_2)$ by defining

$D := \nabla_{1,2} + (-1)^{p+q+1}\delta$ on $\Omega^{p,q}(\mathcal{B}_1, \mathcal{B}_2)$. Then we have $D^2 = 0$ and hence $(\Omega^*(\mathcal{B}_1, \mathcal{B}_2), D)$ is a cochain complex.

We can define another complex $\check{\Omega}^{p,-q} = \Omega^p(M_{1,2}, \mathrm{Hom}_{qu}(E_1, E_2) \otimes \wedge^q \mathcal{N}^*_{1\vee2})$, with two differentials $(\nabla_{1,2}, (-1)^p\check{\delta})$. There is an isomorphism between the two double complexes, as well as their total complexes, by a degree shifting.

Remark 3.7. In fact we should consider those $\phi \in \Omega^*(\mathcal{B}_1, \mathcal{B}_2)$ with decay condition $|\gamma'|^k|\phi(\gamma)| \to 0$ as $|\gamma'| \to \infty$ for all $k \in \mathbb{Z}_{\geq 0}$. We continue to denote this subcomplex by $(\Omega^*(\mathcal{B}_1, \mathcal{B}_2), D)$.

Example 3.8. In the case where $C_{a_1} = \{a_1\}$ and $C_{a_2} = \{a_2\}$ are translations of the zero subtorus C, $M_{1,2}$ is parametrized by the lattice Λ. Let $f(\lambda) \in \Gamma(M_{1,2}, \wedge^0\mathcal{N}_{1\vee2})$, we have

$$\delta(f(\lambda)) = 2\pi i f(\lambda)(\lambda + \tilde{a}_2 - \tilde{a}_1).$$

If $a_1 \neq a_2$, then $\tilde{a}_2 - \tilde{a}_1$ is not in Λ and $\lambda + \tilde{a}_2 - \tilde{a}_1$ is not zero. In this case, $H^0_\delta(\Gamma(\wedge^*\mathcal{N}_{1\vee2}))$ is trivial which means C_{a_1} and C_{a_2} do not intersect in the classical sense. However, if $a_1 = a_2$, then $\tilde{a}_2 - \tilde{a}_1 \in \Lambda$ and we can check that $H^0_\delta(\Gamma(\wedge^*\mathcal{N}_{1\vee2})) = \mathbb{C}$, which means C_{a_1} and C_{a_2} do intersect in the classical sense. Although the spaces of instantons in both cases are the same, the difference in the differential δ helps to distinguish whether the flat branes C_{a_1} and C_{a_2} intersect or not.

When two flat branes are the same, denoted by $\mathcal{B} = (C_a, E)$, we can define a product on $\Omega^*(\mathcal{B}, \mathcal{B})$. We consider a natural map

$$p : M(C_a, C_a) \; {}_{ev_t}\!\times_{ev_s} M(C_a, C_a) \to M(C_a, C_a)$$

defined by joining the paths. More precisely, given a pair of paths (γ_1, γ_2) such that $\gamma_1(1) = \gamma_2(0)$, we let $p(\gamma_1, \gamma_2) = \gamma_1 \circ \gamma_2$ be the unique geodesic path with $\gamma_1 \circ \gamma_2(0) = \gamma_1(0)$ and $(\gamma_1 \circ \gamma_2)' = \gamma_1' + \gamma_2'$. For two $\alpha, \beta \in \Omega^*(\mathcal{B}, \mathcal{B})$, we define

$$\alpha * \beta = p_*(\pi_1^*\alpha \wedge \pi_2^*\beta).$$

In the case where C is just a point in T, we can identify $M(C_a, C_a)$ with the lattice Λ. The product is a convolution product along the lattice. When C is T, it gives the ordinary wedge product. In general, it is a mixture of both. The complex $(\Omega^*(\mathcal{B}, \mathcal{B}), D)$, together with the product $*$, gives a differential graded algebra.

3.1. Fourier transform

In this section, we define the Fourier transform. A check "$\vee$" will be added to notations for objects in T^* to distinguish them from objects in T.

Transform of flat branes

We have to define a transform between flat branes in T and T^* by using the Poincaré line bundle $\mathcal{P}$ and the universal connection $\nabla_{\mathcal{P}}$.

Given a flat brane $\mathcal{B} = (C_a, E)$ in T, without loss of generality, we consider the case that E is a line bundle over C_a, with a flat unitary connection ∇_E.

For each $\check{y} \in T^*$, we have an embedding $C_a \times \{\check{y}\} \to T \times T^*$. We define

$$\check{E}|_{\check{y}} := H^0(E \otimes \mathcal{P}^*|_{C_a \times \{\check{y}\}}, \nabla_E \otimes \nabla_{\mathcal{P}}^*|_{C_a \times \{\check{y}\}}),$$

and we consider the subset $\{\check{y} \in T^* : \check{E}|_{\check{y}} \neq 0\}$ in T^*. It turns out this subset equals to the affine subtorus $\check{C}_{\check{a}}$ in T^*, for a suitable choice of $\check{a}$ (determined by ∇_E).

We consider the bundle $E \otimes \mathcal{P}^*$ over $C_a \times \check{C}_{\check{a}}$. We define $\check{E}|_{\check{y}} = H^0(E \otimes \mathcal{P}^*|_{\check{y}})$. It has a natural metric constructed from the L^2 metric of the space $H^0(E \otimes \mathcal{P}^*|_{\check{y}})$. Parallel transport along paths in $\check{C}_{\check{a}}$ defines a connection $\check{\nabla}_{\check{E}}$ on $\check{E}$. We define $\mathcal{F}(\mathcal{B})$ to be the flat brane $(\check{C}_{\check{a}}, \check{E})$.

This can be seen in local coordinates. Given C_a, we can fix a local coordinate $y = (y_1, \ldots, y_n)$ of V, such that

$$C_{\mathbb{R}} = \{y \in V : y_{k+1} = 0, \cdots, y_n = 0\},$$

and let $\check{y} = (y^1, \ldots, y^n)$ be the corresponding dual coordinates. We can write the flat $U(1)$-connection ∇_E on C_a as the connection 1-form

$$\nabla_{\check{a}} = d - 2\pi i \sum_{j=1}^{k} a^j dy_j$$

with respect to some trivialization 1_E, for some $\check{a} = (a^1, \cdots, a^k) \in C_{\mathbb{R}}^*$.

Fixing a point $\check{y} \in V^*$, the function $e^{-\pi i(\check{y}, y)}$ on $V \times V^*$ descends to a trivialization of $\mathcal{P}|_{T \times \{\check{y}\}}$, with corresponding connection 1-form given by $d - 2\pi i(\check{y}, dy)$. Then we have

$$\nabla_{\check{a}} \otimes \nabla_{\mathcal{P}}^*|_{C_a \times \{\check{y}\}} = d + 2\pi i \sum_{j=1}^{k} (y^j - a^j) dy_j,$$

it has nonzero flat section if and only if

$$f(y_1, \cdots, y_k) = \exp\left(- 2\pi i \sum_{j=1}^{k} (y^j - a^j) y_j \right)$$

defines a function on C. This is equivalent to the condition $\check{y} - \check{a} \in C_{\mathbb{Z}}^*$.

We can verify that, for a suitable choice of $a \in V/C_{\mathbb{R}}$, the connection $\check{\nabla}_{\check{E}}$ can be expressed in the form

$$\check{\nabla}_a = d - 2\pi i \sum_{j=k+1}^{n} a_j dy^j,$$

with respect to some trivialization of $\check{E}$.

In general, $\mathcal{B} = (C_a, E)$ is a flat brane in T with E a flat $U(r)$-bundle. Since C_a itself is a torus, (E, ∇) can always be decomposed into a direct sum $\bigoplus_{i=1}^{r}(E_i, \nabla_i)$. Thus, transformation of $\mathcal{B}$ reduces to transformation of (C_a, E_i) and gives r flat branes in T^*.

Finally, we define a tautological section ϕ of $E^* \otimes \check{E} \otimes \mathcal{P}$ over $C_a \times \check{C}_{\check{a}}$ which is used in the transformation of the quantum intersection complex. Over a point $(y, \check{y})$, there is a natural pairing

$$E^*|_y \otimes H^0(E \otimes \mathcal{P}^*|_{\check{y}}) \otimes \mathcal{P}|_{(y,\check{y})} \to \mathbb{C}.$$

This gives an identification of the bundle $E^* \otimes \check{E} \otimes \mathcal{P}$ with the trivial bundle and determines a section ϕ which is identified to the constant function 1. Notice that ϕ is flat with respect to the connection of $E^* \otimes \check{E} \otimes \mathcal{P}$.

The inverse transform $\mathcal{F}^{-1}$ can be defined in a similar way using the dual Poincaré line bundle $\mathcal{P}^*$, and we have

Proposition 3.9. $\mathcal{F}^{-1} \circ \mathcal{F}(\mathcal{B}) = \mathcal{B}$.

Remark 3.10. When we carry out the inverse transform $\mathcal{F}^{-1}$ on $\mathcal{F}(\mathcal{B}) = (\check{C}_{\check{a}}, \check{E})$, we get a line bundle $\check{\check{E}}$ over C_a which is defined by $\check{\check{E}}|_y := H^0(\check{E} \otimes \mathcal{P}|_{\{y\}\times \check{C}_{\check{a}}}, \check{\nabla}_{\check{E}} \otimes \nabla_{\mathcal{P}}|_{\{y\}\times \check{C}_{\check{a}}})$. The tautological section ϕ gives an identification of E and $\check{\check{E}}$.

Transform of quantum intersection complex

Given flat branes $\mathcal{B}_i = (C_{i,a_i}, E_i)$ in T, for $i = 1, 2$, we are going to define a Fourier transform

$$\mathcal{F} : \Omega^*(\mathcal{B}_1, \mathcal{B}_2) \to \Omega^*(\mathcal{F}(\mathcal{B}_1), \mathcal{F}(\mathcal{B}_2)).$$

We write $\mathcal{F}(\mathcal{B}_i) = (\check{C}_{i,\check{a}_i}, \check{E}_i)$. For simplicity, we work on the case of line bundles.

Let us consider the product of spaces $M_{1,2} \times \check{M}_{1,2}$ with two natural projections π and $\check{\pi}$ to $M_{1,2}$ and $\check{M}_{1,2}$ respectively. As analogue to the Fourier-Mukai transform, our transform is established by defining some universal objects over the space $M_{1,2} \times \check{M}_{1,2}$.

First, we define a tautological section $\phi_{\mathcal{F}}$ of the bundle

$$\pi^* \mathrm{Hom}_{qu}(E_1, E_2)^* \otimes \check{\pi}^* \mathrm{Hom}_{qu}(\check{E}_1, \check{E}_2)$$

on $M_{1,2} \times \check{M}_{1,2}$. The product of evaluation maps gives

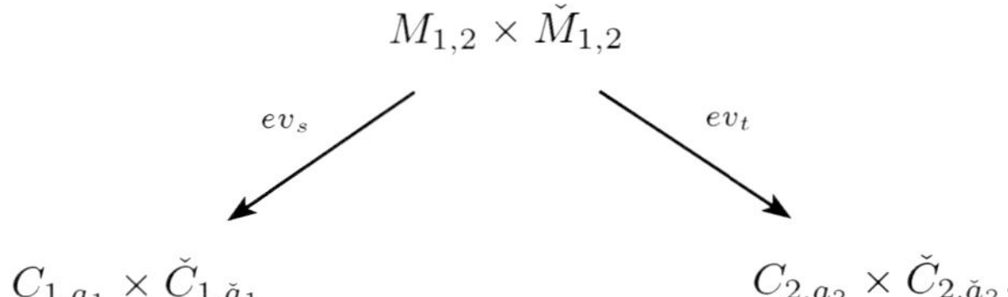

$$M_{1,2} \times \check{M}_{1,2}$$

$$ev_s \swarrow \qquad\qquad \searrow ev_t$$

$$C_{1,a_1} \times \check{C}_{1,\check{a}_1} \qquad\qquad C_{2,a_2} \times \check{C}_{2,\check{a}_2}.$$

We also have two tautological sections ϕ_i of $E_i^* \otimes \check{E}_i \otimes \mathcal{P}$ on $C_{i,a_i} \times \check{C}_{i,\check{a}_i}$, for $i = 1, 2$. $(ev_s)^* \phi_1^* \otimes (ev_t)^* \phi_2$ is a section of the bundle

$$\pi^* \operatorname{Hom}_{qu}(E_1, E_2)^* \otimes \check{\pi}^* \operatorname{Hom}_{qu}(\check{E}_1, \check{E}_2) \otimes (ev_s)^* \mathcal{P}^* \otimes (ev_t)^* \mathcal{P}.$$

Each point $(\gamma, \check{\gamma}) \in M_{1,2} \times \check{M}_{1,2}$ gives a path in $T \times T^*$ and parallel transport along it gives a natural pairing $((ev_s)^* \mathcal{P}^* \otimes (ev_t)^* \mathcal{P})|_{(\gamma,\check{\gamma})} \to \mathbb{C}$. We define $\phi_{\mathcal{F}}$ to be the section $(ev_s)^* \phi_1^* \otimes (ev_t)^* \phi_2$ after identifying $(ev_s)^* \mathcal{P}^* \otimes (ev_t)^* \mathcal{P}$ with $\mathbb{C}$ using the natural pairing.

Next, we define two canonical forms

$$\begin{aligned}
G_1 &\in \pi^* \Omega^1(M_{1,2}, \wedge^1 \mathcal{N}_{1\vee 2}) \\
G_2 &\in \check{\pi}^* \Omega^1(\check{M}_{1,2}, \wedge^1 \mathcal{N}_{1\vee 2}).
\end{aligned}$$

Through the evaluation map $ev_s : M_{1,2} \to C_{1,a_1}$, we identify $T_\gamma M_{1,2}$ with the subspace $(T_{\gamma(0)} C_{1,a_1} \cap P_\gamma^{-1} T_{\gamma(1)} C_{2,a_2})$ of $T_{\gamma(0)} T$. At the path $(\gamma, \check{\gamma})$, using the natural pairing on $T_{\gamma(0)} T \otimes T_{\check{\gamma}(0)} \check{T}$, we obtain an isomorphism between $T_\gamma^* M_{1,2}$ and $\mathcal{N}_{1\vee 2}|_{\check{\gamma}}$. Hence we have two natural sections G_1, G_2 of the bundle

$$\pi^* T^* M_{1,2} \otimes \check{\pi}^* \check{\mathcal{N}}_{1\vee 2} \oplus \check{\pi}^* T^* \check{M}_{1,2} \otimes \pi^* \mathcal{N}_{1\vee 2}$$

given by restrictions of the metric and the dual metric to corresponding subspaces.

The universal section $\phi_{\mathcal{F}} \otimes e^{G_1 + G_2}$ serves as the kernel function for the Fourier transform. Given $\alpha \in \Omega^*(\mathcal{B}_1, \mathcal{B}_2)$, we define

$$\mathcal{F}(\alpha) = (-1)^{m+\check{m}} \check{\pi}_* (\pi^* \alpha \wedge (\phi_{\mathcal{F}} \otimes e^{G_1 + G_2})),$$

where m and $\check{m}$ are dimensions of $M_{1,2}$ and $\check{M}_{1,2}$ respectively.

Here we choose unit length trivialization of the top exterior power of N_{1v2} with R using the metric, together with an orientation of $M_{1,2}$ to obtain an integration

$$\int : \Omega^m(M_{1,2}, \wedge^{\check{m}} \mathcal{N}_{1\vee 2}) \to \mathbb{C}.$$

In the case that $\mathcal{B}_1 = \mathcal{B}_2 = T$, equipped with the trivial connection, $\check{\mathcal{B}}$ is the trivial subtorus without affine translation. In this case, we recover the ordinary Fourier transform.

Example 3.11. For simplicity, we let $T = \mathbb{S}^1 = \mathbb{R}/\mathbb{Z}$ be the unit circle with coordinate $\theta \in \mathbb{R}$, and $\check{\theta} \in \mathbb{R}^*$ be the corresponding dual coordinate. We write $\Lambda \simeq \mathbb{Z}$ for the lattice and $\Lambda^* \simeq \mathbb{Z}^*$ for the dual lattice. $\check{\lambda} \in \mathbb{Z}^*$ parametrizes the loop which is a line segment from $\check{0}$ to $\check{\lambda}$ in $\mathbb{R}^*$. The universal connection 1-form is given by $\nabla_{\mathcal{P}} = d + \pi i(\theta d\check{\theta} - \check{\theta} d\theta)$ on $\mathbb{R} \times \mathbb{R}^*$.

We take $C_1 = C_2 = T$, equipped with the trivial connection on the trivial bundle. Then $\check{C}_1 = \check{C}_2 = \check{0}$ is the origin. We parametrize $M_{1,2}$ by T and $\check{M}_{1,2}$ by Λ^*. We only need to compute the parallel transport along the path $\check{\gamma}_{\check{\lambda}}$. Parallel transport of $\mathcal{P}$ along $\check{\gamma}_{\check{\lambda}}$ maps 1 to $e^{-\pi i \check{\lambda}\theta} \cdot 1$ on the universal cover, taking the quotient and changing the trivialization results in a multiplication by $e^{-2\pi i \check{\lambda}\theta}$. Hence the tautological section $\phi_{\mathcal{F}}$ is given by $e^{2\pi i \check{\lambda}\theta}$ and we have $\mathcal{F}(f d\theta)(\check{\lambda}) = -\int_T f(\theta) e^{2\pi i \check{\lambda}\theta} d\theta$, under the natural invariant basis of the bundles.

We give three more examples to illustrate the situation.

Example 3.12. In this example, we consider the standard torus $T = \mathbb{R}^n/\mathbb{Z}^n$, equipped with the standard metric. We use $y = (y_1, \ldots, y_n) \in \mathbb{R}^n$ as coordinate and let $\check{y} = (y^1, \ldots, y^n) \in (\mathbb{R}^n)^*$ be the corresponding dual coordinate. We consider two flat branes given by

$$\mathcal{B}_i = \left(\tilde{C}_i = \{y : y_k = 0 \ \ for \ \ k \in I_i\}, \ E_i = \mathbb{C} \cdot 1_{E_i}, \ \nabla_{\check{a}_i} = d - 2\pi i \sum_{j \notin I_i} a_i^j dy_j \right),$$

with $\tilde{C}_i$ are some liftings of C_i to V. The corresponding dual branes are given by

$$\check{\mathcal{B}}_i = \left(\tilde{\check{C}}_{i,\check{a}_i} = \{y : y^k = a_i^k \ \ for \ \ k \notin I_i\}, \ \check{E}_i = \mathbb{C} \cdot \check{1}_{\check{E}_i}, \ \check{\nabla} = d \right),$$

for some trivialization $\check{1}_{\check{E}_i}$. We also denote by $\check{a}_i \in \check{C}_{i,\mathbb{R}}^\perp$ some fixed liftings of the affine translations if there is no confusion.

In this case, we give explicit parametrization of the spaces $M_{1,2}$ and $\check{M}_{1,2}$. We denote by $\check{C}_{1\vee 2,\mathbb{Z}}^\perp$ the lattice $(\check{C}_{1,\mathbb{R}} + \check{C}_{2,\mathbb{R}})^\perp \cap \Lambda$. There is an isomorphism

$$(\check{C}_1 \cap \check{C}_2) \times \check{C}_{1\vee 2,\mathbb{Z}}^\perp \to \check{M}_{1,2},$$

given by sending $(\check{y}, \check{\lambda}) \in (\check{C}_1 \cap \check{C}_2) \times \check{C}_{1\vee 2,\mathbb{Z}}^\perp$ to a path $\check{\gamma}_{\check{y},\check{\lambda}}$ with $\check{\gamma}_{\check{y},\check{\lambda}}(0) = \check{y} + \check{a}_1 + \check{a}_2^T$, $\check{\gamma}'_{\check{y},\check{\lambda}} = \check{\lambda} + (\check{a}_2 - \check{a}_1)^\perp$ and $\check{\gamma}_{\check{y},\check{\lambda}}(1) = \check{y} + \check{a}_2 + \check{a}_1^T$, where $\check{a}_1^T$ is the projection of $\check{a}_1$ to $\check{C}_{2,\mathbb{R}}$ (similarly for $\check{a}_2$) and $(\check{a}_2 - \check{a}_1)^\perp$ is the projection to $\check{\mathcal{N}}_{1\vee 2} \simeq \check{C}_{1\vee 2,\mathbb{R}}^\perp$. The space $M_{1,2} \simeq (C_1 \cap C_2) \times C_{1\vee 2,\mathbb{Z}}^\perp$ is parametrized in similar way.

We give an explicit description of the tautological section $\phi_{\mathcal{F}}$. In $T \times T^*$, consider a path $(\gamma_{y,\lambda}, \check{\gamma}_{\check{y},\check{\lambda}})$. In order to compute the sections ϕ_i, we choose liftings $\tilde{C}_1 = C_{1,\mathbb{R}}$ and $\tilde{C}_{2,\lambda} = C_{2,\mathbb{R}} + \lambda$ of C_1 and C_2, liftings $\tilde{\check{C}}_{1,\check{a}_1} = \check{C}_{1,\mathbb{R}} + \check{a}_1$ and $\tilde{\check{C}}_{2,\check{a}_2+\check{\lambda}} = \check{C}_{2,\mathbb{R}} + \check{a}_2 + \check{\lambda}$ of $\check{C}_{1,\check{a}_1}$ and $\check{C}_{2,\check{a}_2}$ respectively. The functions $e^{-\pi i(\check{a}_1, y)}$ on $\tilde{C}_1 \times \tilde{\check{C}}_{1,\check{a}_1}$ and $e^{-\pi i[(\check{a}_2+\check{\lambda},y)-(\lambda,\check{y})]}$ on

$\tilde{C}_2 \times \tilde{\check{C}}_{2,\check{a}_2+\check{\lambda}}$ descend to trivializations e_1 and e_2 of $\mathcal{P}|_{C_1 \times \check{C}_{1,\check{a}_1}}$ and $\mathcal{P}|_{C_2 \times \check{C}_{2,\check{a}_2}}$ respectively.

Similar to the previous example, we see that the parallel transport of $\mathcal{P}$ on $V \times V^*$ along $(\gamma_{y,\lambda}, \check{\gamma}_{\check{y},\check{\lambda}})$ is multiplication by a factor $e^{-\pi i[(\check{\lambda}+(\check{a}_2-\check{a}_1)^\perp, y)-(\lambda,\check{y})]}$. It follows that the parallel transport maps e_1 to e_2.

The connection forms of the Poincaré line bundle over $C_1 \times \check{C}_{1,\check{a}_1}$ and $C_2 \times \check{C}_{2,\check{a}_2}$ (corresponding to the trivializations e_1 and e_2) are written as $d - 2\pi i(\check{a}_1, dy)$ and $d + 2\pi i[(\lambda, d\check{y}) - (\check{a}_2 + \check{\lambda}, dy)]$ respectively. We choose the trivializations of $\check{E}_1$ and $\check{E}_2$ given by $1_{E_1} \otimes e_1^*$ and $e^{2\pi i[(\lambda,\check{y})-(\check{\lambda},y)]}1_{E_2} \otimes e_2^*$, denoted by $\check{1}_{\check{E}_1}$ and $\check{1}_{\check{E}_2}$ respectively. Notice that $\check{1}_{\check{E}_i}$ do not depend on λ and $\check{\lambda}$. As a result, we have $\phi_1 = 1_{E_1}^* \otimes \check{1}_{\check{E}_1} \otimes e_1$ and $\phi_2 = e^{2\pi i[(\check{\lambda},y)-(\lambda,\check{y})]}1_{E_2}^* \otimes \check{1}_{\check{E}_2} \otimes e_2$. Hence we have

$$\phi_{\mathcal{F}} = e^{2\pi i[(\check{\lambda},y)-(\lambda,\check{y})]}1_{E_1} \otimes 1_{E_2}^* \otimes \check{1}_{\check{E}_1}^* \otimes \check{1}_{\check{E}_2}.$$

Next, we give the expression of the canonical forms. Under the identification of $T^*M_{1,2} \simeq \check{\mathcal{N}}_{1\vee2}$, dy_k is identified with $\frac{\partial}{\partial y^k}$, for $k \in I_1 \cap I_2$. Similarly, dy^k is identified with $\frac{\partial}{\partial y_k}$, for $k \in \bar{I}_1 \cap \bar{I}_2$, where $\bar{I}_i = \{1, \ldots, n\} - I_i$. Then we have

$$G_1 + G_2 = \sum_{k \in I_1 \cap I_2} dy_k \otimes \frac{\partial}{\partial y^k} - \sum_{l \in \bar{I}_1 \cap \bar{I}_2} \frac{\partial}{\partial y_l} \otimes dy^l.$$

Suppose we have $\alpha \in \Omega^*(B_1, B_2)$ written in the form

$$\alpha(y, \lambda) = f(y, \lambda)dy_J \otimes \frac{\partial}{\partial y_K}.$$

Then we will have

$$\mathcal{F}(\alpha)(\check{y}, \check{\lambda})$$

$$= \left(\int_{C_1 \cap C_2} (\sum_\lambda f(y, \lambda)e^{-2\pi i(\lambda,\check{y})})e^{2\pi i(\check{\lambda},y)}dy_{I_1 \cap I_2} \otimes \frac{\partial}{\partial y_{\bar{I}_1 \cap \bar{I}_2}} \right) dy^{(\bar{I}_1 \cap \bar{I}_2)-K} \otimes \frac{\partial}{\partial y^{(I_1 \cap I_2)-J}},$$

expressed in terms of the corresponding trivialization, up to a sign.

As seen from the above example, the Fourier transform is a homomorphism

$$\mathcal{F} : \Omega^{p,q}(\mathcal{B}_1, \mathcal{B}_2) \to \Omega^{\check{m}-q,m-p}(\check{\mathcal{B}}_1, \check{\mathcal{B}}_2).$$

We can use the Hodge star operator over the trivial bundles $\mathcal{N}_{1\vee2}$ and $\check{\mathcal{N}}_{1\vee2}$ to turn $\mathcal{F}$ into a homomorphism

$$\hat{\mathcal{F}} : \Omega^{p,q}(\mathcal{B}_1, \mathcal{B}_2) \to \Omega^{q,p}(\check{\mathcal{B}}_1, \check{\mathcal{B}}_2).$$

After doing that, we have

Proposition 3.13. $\hat{\mathcal{F}} : \Omega^*(\mathcal{B}_1, \mathcal{B}_2) \to \Omega^*(\check{\mathcal{B}}_1, \check{\mathcal{B}}_2)$ *is a chain map.*

Remark 3.14. Indeed it is more convenient to write down the transform using the complex $\check{\Omega}^*(\mathcal{B}_1, \mathcal{B}_2)$. The advantage of this is that the transform preserves the differential without involving the star operator and the metric tensor. Let I_1 be a section of the bundle $T^*M_{1,2} \otimes \check{\mathcal{N}}^*_{1\vee 2}$ defined using the natural pairing between $\check{\mathcal{N}}_{1\vee 2}$ and $TM_{1,2}$, and let I_2 defined similarly. The transform $\check{\mathcal{F}}$ is defined by

$$\check{\mathcal{F}}(\alpha) = (-1)^{m+\check{m}} \check{\pi}_* (\pi^*\alpha \wedge (\phi_{\mathcal{F}} \otimes e^{I_1 + I_2})).$$

Example 3.15. In this example, we consider an example where the two branes are the same. Let $\mathcal{B}$ be a brane in T^2 with a lifting $\tilde{C}$ of the torus C given by

$$(\tilde{C} = \{y : y_2 - 2y_1 = 0\}, \ E = \mathbb{C} \cdot 1_E, \ \nabla = d).$$

The lifting $\tilde{\check{C}}$ of the dual torus with corresponding dual connection is given by

$$(\tilde{\check{C}} = \{\check{y} : 2y^2 + y^1 = 0\}, \ \check{E} = \mathbb{C} \cdot \check{1}_{\check{E}}, \ \check{\nabla} = d).$$

In this case, we use explicit parametrizations of M and $\check{M}$, with coordinates (y_1, λ) and $(y^2, \check{\lambda})$. A path $\gamma_{y_1, \lambda}$ with coordinate (y_1, λ) has $\gamma(0) = (y_1, 2y_1)$ and $\gamma'_{y_1, \lambda} = \lambda(\frac{-2}{5}, \frac{1}{5})$. A path $\check{\gamma}_{y^2, \check{\lambda}}$ has $\check{\gamma}_{y^2, \check{\lambda}}(0) = (-2y^2, y^2)$ and $\check{\gamma}'_{y^2, \check{\lambda}} = \lambda(\frac{1}{5}, \frac{2}{5})$. For example, $\gamma'_{0,3}$ is shown in the following figure.

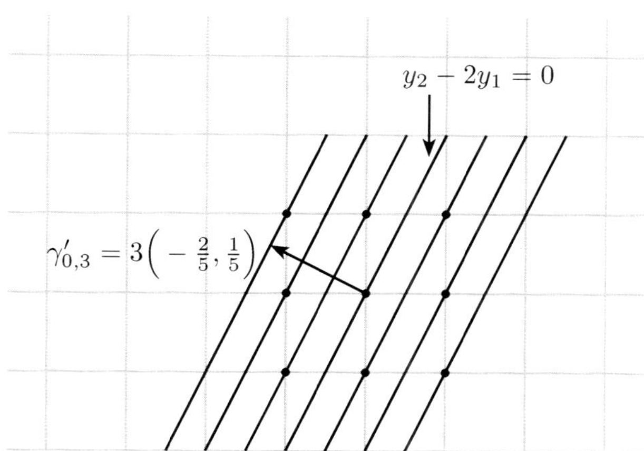

We consider a path $(\gamma_{y_1, \lambda}, \check{\gamma}_{y^2, \check{\lambda}})$ in $T \times T^*$. The constant function 1 descends to a trivialization e_1 of the Poincaré bundle over $C \times \check{C}$. We have $\phi_1 = 1_E \otimes \check{1}_{\check{E}} \otimes e_1$.

Then we use the liftings $\tilde{C}_{(0,\lambda)} = C_{\mathbb{R}} + (0,\lambda)$ and $\check{\tilde{C}}_{(\check{\lambda},0)} = \check{C}_{\mathbb{R}} + (\check{\lambda},0)$ to compute ϕ_2. Note that the function $e^{\pi i(\lambda \check{y}^2 - \check{\lambda} y_1)}$ on $\tilde{C}_{(0,\lambda)} \times \check{\tilde{C}}_{(\check{\lambda},0)}$ descends to a trivialization e_2 of $\mathcal{P}|_{C_{(0,\lambda)} \times \check{C}_{(\check{\lambda},0)}}$, with corresponding connection form given by $d + 2\pi i(\lambda d\check{y}^2 - \check{\lambda} dy_1)$. We have $\phi_2 = e^{2\pi i(\check{\lambda} y_1 - \lambda y^2)} 1_E \otimes \check{1}_{\check{E}} \otimes e_2$. Parallel transport along $(\gamma_{y_1,\lambda}, \check{\gamma}_{y^2,\check{\lambda}})$ identifies e_1 to e_2. Thus we obtain the tautological section given by

$$\phi_{\mathcal{F}} = e^{2\pi i[(\check{\lambda},y_1) - (\lambda,y^2)]} 1_E \otimes 1_E^* \otimes \check{1}_{\check{E}}^* \otimes \check{1}_{\check{E}},$$

and the Fourier transform for a function $f(y_1, \lambda)$ is given by

$$\hat{\mathcal{F}}(f)(y^2, \check{\lambda}) = \sum_{\lambda \in \mathbb{Z}} \left(\int_C f(y_1, \lambda) e^{2\pi i[(\check{\lambda},y_1) - (\lambda,y^2)]} \right),$$

with respect to the chosen basis.

Example 3.16. We take the standard 2-dimensional torus T^2 in this example. We use same notations as last example. The branes $\mathcal{B}_i$ have liftings given by

$$(\tilde{C}_1 = \{y : y_2 = 0\}, \ E_1 = \mathbb{C} \cdot 1_{E_1}, \ \nabla_1 = d)$$
$$(\tilde{C}_2 = \{y : y_2 - ny_1 = 0\}, \ E_2 = \mathbb{C} \cdot 1_{E_2}, \ \nabla_2 = d).$$

The corresponding dual branes will have the liftings given by

$$(\check{\tilde{C}}_1 = \{\check{y} : y^1 = 0\}, \ \check{E}_1 = \mathbb{C} \cdot \check{1}_{\check{E}_1}, \ \check{\nabla}_1 = d)$$
$$(\check{\tilde{C}}_2 = \{\check{y} : ny^2 + y^1 = 0\}, \ \check{E}_2 = \mathbb{C} \cdot \check{1}_{\check{E}_2}, \ \check{\nabla}_2 = d).$$

for some $n \in \mathbb{Z}_{>0}$.

In this case, $M_{1,2}$ and $\check{M}_{1,2}$ consist of a finite number of constant paths. We denote $\gamma_i = (\frac{i}{n}, i)$ and $\check{\gamma}_i = (i, -\frac{i}{n})$, for $i = 0, \cdots, n-1$. The constant function 1 on $\tilde{C}_2 \times \check{\tilde{C}}_2$ descends to a trivialization e_2 of $\mathcal{P}|_{C_2 \times \check{C}_2}$. To compute the $\phi_{\mathcal{F}}$ at $(\gamma_k, \check{\gamma}_j)$, we choose $\tilde{C}_{1,(0,k)} = \tilde{C}_1 + (0,k)$ and $\check{\tilde{C}}_{1,(j,0)} = \check{\tilde{C}}_1 + (j,0)$ as liftings of the subtori on the universal cover, as shown in the figure.

18

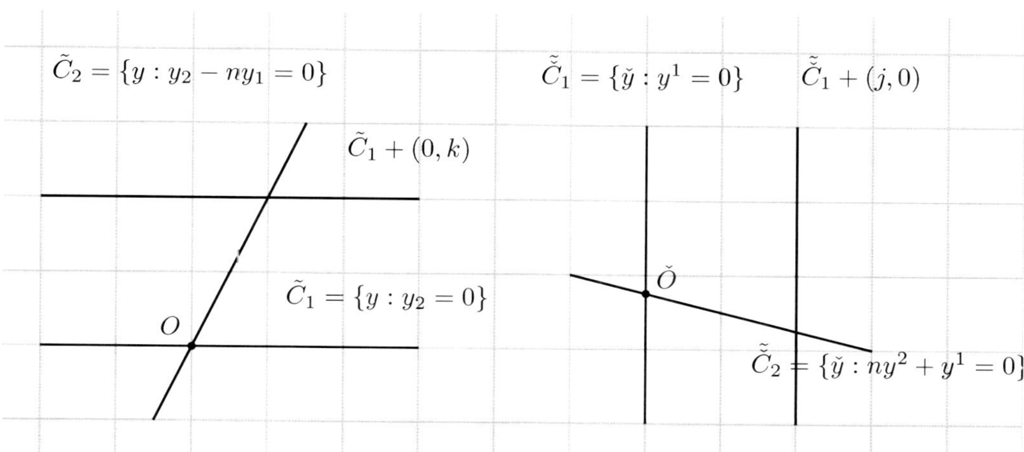

The function $e^{\pi i(ky^2 - jy_1)}$ on $\tilde{C}_{1,(0,k)} \times \tilde{\tilde{C}}_{1,(j,0)}$ descends to a trivialization e_1 of $\mathcal{P}|_{C_1 \times \check{C}_1}$. Hence, we get $\phi_1 = e^{2\pi i(jy_1 - ky^2)} 1^*_{E_1} \otimes \check{1}_{\check{E}_1} \otimes e_1$. At the point $(\gamma_k, \check{\gamma}_j)$, we compute that $\phi_{\mathcal{F}} = \left(e^{-2\pi i \frac{jk}{n}} \right)_{j,k}$, which is an invertible matrix transforming functions on $M_{1,2}$ to functions on $\check{M}_{1,2}$.

The general case is a mixture of the above situations and we claim that for $\hat{\mathcal{F}}^{-1}$ defined similarly, we have

Proposition 3.17.

$$\hat{\mathcal{F}}^{-1} \circ \hat{\mathcal{F}} = id$$

up to a constant.

4. Mirror symmetry without corrections

In this section, we will have a brief review of certain mirror symmetry phenomena for semi-flat Calabi-Yau manifolds [11, 13]. In this case, mirror symmetry is T-duality without any modifications (or so-called quantum corrections). In fact, by performing the classical Fourier transform

$$\mathcal{F}_{cl} : \Omega^*(T) \to \Omega^*(T^*)$$

to each torus fiber, we will see the interchange of complex and symplectic structures on mirror manifolds. Furthermore, we will give a taste of how the Fourier transform on flat branes defined in the last section establishes the correspondence between A-branes and B-branes.

4.1. Semi-flat mirror Calabi-Yau manifolds

Let $N \cong \mathbb{Z}^n$ be a lattice and $M = \mathrm{Hom}(N, \mathbb{Z})$ denote the dual lattice of N. We denote by $N_{\mathbb{R}} = N \otimes_{\mathbb{Z}} \mathbb{R}$ and $M_{\mathbb{R}} = M \otimes_{\mathbb{Z}} \mathbb{R}$ the real vector spaces spanned by N and M respectively. Let $B \subset M_{\mathbb{R}}$ be a convex domain. (In general, one may consider an affine manifold.) We can construct two (non-compact) Calabi-Yau manifolds X and $\check{X}$ from the tangent and cotangent bundles of B, they are called mirror manifolds.

Construction of $\check{X}$

Firstly, the tangent bundle $TB = B \times iM_{\mathbb{R}}$ is naturally a complex manifold with complex coordinates $z^j = b^j + iy^j$'s, where b^j's and y^j's are the base coordinates on B and corresponding fiber coordinates on $M_{\mathbb{R}}$. TB is equipped with the standard holomorphic volume form $\Omega_{TB} = dz^1 \wedge \cdots \wedge dz^n$. Taking quotient of TB by the lattice $iM \subset iM_{\mathbb{R}}$, each fiber is compactified to a torus. We denote

$$\check{X} := TB/iM = B \times iT_M,$$

where T_M denotes the torus $M_{\mathbb{R}}/M$ and $\check{p} : \check{X} \to B$ is a torus fibration over B. Then the holomorphic volume form Ω_{TB} on TB descends to give the holomorphic volume form $\Omega_{\check{X}} = dz^1 \wedge \cdots \wedge dz^n$ on $\check{X}$.

Furthermore, if ϕ is an elliptic solution of the real Monge-Ampére equation

$$\det \left(\frac{\partial^2 \phi}{\partial b^j \partial b^k} \right) = constant,$$

then the Kähler form

$$\omega_{\check{X}} = 2i\partial\bar{\partial}\phi = \sum_{j,k} \phi_{jk} db^j \wedge dy^k,$$

where ϕ_{jk} is defined to be $\frac{\partial^2 \phi}{\partial b^j \partial b^k}$, determines a Calabi-Yau metric on $\check{X}$. Then $\check{p} : \check{X} \to B$ becomes a special Lagrangian torus fibration, which is called the SYZ fibration.

Construction of X

Next, we consider the cotangent bundle T^*B of B, we have $T^*B = B \times iN_{\mathbb{R}}$. T^*B carries the standard symplectic form $\omega_{T^*B} = \sum_{j=1}^{n} db^j \wedge dy_j$, where y_j's are fiber coordinates on $N_{\mathbb{R}}$. Taking the quotient of T^*B by the dual lattice $iN \subset iN_{\mathbb{R}}$, each fiber is compactified to be a torus, denoted by

$$X := T^*B/iN = B \times iT_N,$$

where T_N denotes the torus $N_{\mathbb{R}}/N$. Note that T_N is the dual torus of T_M and $p : X \to B$ is the dual torus fibration of $\check{p}$ over B. Then the symplectic form ω_{T^*B} descends to give the symplectic form $\omega_X = \sum_{j=1}^{n} db^j \wedge dy_j$ on X.

By considering the metric

$$g_X = \sum_{j,k} (\phi_{jk} db^j \otimes db^k + \phi^{jk} dy_j \otimes dy_k),$$

where (ϕ^{jk}) is the inverse matrix of ϕ_{jk}, we obtain a complex structure on X with complex coordinates given by $dz_j = \sum_{k=1}^{n} \phi_{jk} db^k + i dy_j$. Then we have a natural holomorphic volume form which is

$$\Omega_X = dz_1 \wedge \cdots \wedge dz_n = \bigwedge_{j=1}^{n} \left(\sum_{k=1}^{n} \phi_{jk} db^k + i dy_j \right).$$

Also $p : X \to B$ beccmes a special Lagrangian torus fibration.

4.2. Semi-flat SYZ transform

In this section, we will discuss how the geometric structures of the mirror manifolds $\check{X}$ and X are transformed to each other by performing fiberwise Fourier transformion.

Given a point $x \in B$, $\check{p}^{-1}(x)$ is the torus T_M and $p^{-1}(x)$ is the torus T_N which is the dual torus of T_M. Using the construction in section 2.2, we define the trivial line bundle $\tilde{\mathcal{P}} = N_{\mathbb{R}} \times M_{\mathbb{R}} \times \mathbb{C}$ over $N_{\mathbb{R}} \times M_{\mathbb{R}}$, equipped with the connection 1-form

$$\nabla_{\mathcal{P}} := d + \pi i \sum_{j=1}^{n} (y_j dy^j - y^j dy_j).$$

Taking quotient by the lattice $N \times M$, we obtain the Poincaré line bundle $\mathcal{P}$. The curvature of $\mathcal{P}$ is

$$F = 2\pi i \sum_{j=1}^{n} dy_j \wedge dy^j.$$

We perform this for each fiber torus and dual torus and combine these Poincaré line bundles together to get a relative version of the above picture.

Let $X \times_B \check{X} = B \times i(T_N \times T_M)$ be the fiber product of the fibrations $p : X \to B$ and $\check{p} : \check{X} \to B$, with two natural projections $\pi : X \times_B \check{X} \to X$ and $\check{\pi} : X \times_B \check{X} \to \check{X}$ respectively. We use $\mathcal{P}$ and $F = 2\pi i \sum_{j=1}^{n} dy_j \wedge dy^j \in \Omega^2(X \times_B \check{X})$ again to denote the fiberwise Poincaré line bundle and curvature two form respectively.

Definition 4.1. The semi-flat SYZ transform $\mathcal{F} : \Omega^*(X) \to \Omega^*(\check{X})$ is defined by

$$\mathcal{F}(\alpha) = \check{\pi}_*(\pi^* \alpha \wedge e^{\frac{i}{2\pi} F}) = \int_{T_N} \pi^* \alpha \wedge e^{\frac{i}{2\pi} F}.$$

Remark 4.2. T_N and T_M here should be identified with T and T^* in section 2.

One significant property of this fiberwise Fourier transform is transforming the symplectic structure on X to the complex structure on $\check{X}$ in the sense of the following proposition.

Proposition 4.3.

$$\mathcal{F}(e^{\omega_X}) = \Omega_{\check{X}}.$$

Furthermore, we may define the inverse Fourier transform $\mathcal{F}^{-1} : \Omega^*(\check{X}) \to \Omega^*(X)$ by

$$\mathcal{F}^{-1}(\check{\alpha}) = i^{-n}\pi_*(\check{\pi}^*\check{\alpha} \wedge e^{-\frac{i}{2\pi}F}) = i^{-n}\int_{T_M} \check{\pi}^*\check{\alpha} \wedge e^{-\frac{i}{2\pi}F}.$$

Then, we transform the complex structure on $\check{X}$ to the symplectic structure on X as in the previous proposition.

Proposition 4.4.

$$\mathcal{F}^{-1}(\Omega_{\check{X}}) = e^{\omega_X}.$$

As expected, one can also check that $\mathcal{F}^{-1}(e^{\omega_{\check{X}}}) = \Omega_X$ and $\mathcal{F}(\Omega_X) = e^{\omega_{\check{X}}}$. The propositions above can be proved by direct calculation [9]. For further discussions on the transform and related materials, we refer readers to [11].

4.3. Fourier transform and semi-flat branes

As described in section 3, for A-branes and B-branes which can be regarded as families of flat branes on tori, it is predicted that the mirror correspondence is a fiberwise Fourier-type transform. In this section, we restrict our attention to the case that all bundles are line bundles and establish this correspondence between special classes of A-branes on X and B-branes on $\check{X}$ using the construction given in section 3.

Semi-flat A-branes

As mentioned in section 1, an A-brane is a pair (L, E) where L is a Lagrangian submanifold and E is a flat unitary complex line bundle (in general, vector bundle) on L. As our space X admits a torus fibration $p : X \to B$, we restrict our attention to those semi-flat A-branes which are compatible with the torus fibration structure $p : X \to B$. More precisely, a semi-flat Lagrangian L is a submanifold of $p^{-1}(B_L)$, where B_L is a rational affine submanifold of B, such that $p|_L : L \to B_L$ is a torus fiber bundle over B_L, and the fiber of $p|_L$ at every point b is an affine subtorus of $p^{-1}(b)$.

Definition 4.5. An A-brane (L, E) on X is called semi-flat if L is a semi-flat Lagrangian in X.

As analogue to section 3, we study the quantum intersection theory between two semi-flat A-branes (L_i, E_i), $i = 1, 2$, by considering the space of fiberwise minimal geodesic paths from L_1 to L_2. This space, denoted by $M(L_1, L_2)$ (or simply $M_{1,2}$ if there is no confusion), is a finite dimensional manifold. There are two evaluation maps

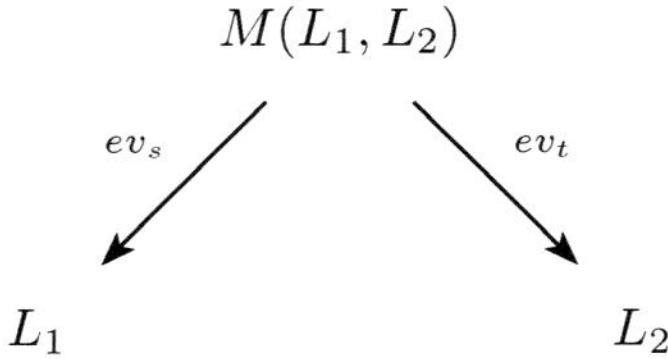

and $p \circ ev_s = p \circ ev_t : M(L_1, L_2) \to B_{L_1} \cap B_{L_2}$ is a fiber bundle, where the fiber over a point b is the space of instantons $M_b(L_1 \cap p^{-1}(b), L_2 \cap p^{-1}(b))$ defined in section 3. We combine $\mathcal{N}_{1 \vee 2, b}$'s on $M_b(L_1 \cap p^{-1}(b), L_2 \cap p^{-1}(b))$ along the base $B_{L_1} \cap B_{L_2}$ to obtain a relative normal bundle of $M_{1,2}$, again denoted by $\mathcal{N}_{1 \vee 2}$, equipped with a connection induced from the metric. The natural section $s \in \Gamma(M_{1,2}, \mathcal{N}_{1 \vee 2})$ is defined similarly. We define $\Omega_A^*((L_1, E_1), (L_2, E_2))$, with the operator $D = \nabla_{1,2} + (-1)^{p+q+1}\delta$ as in section 3. It is not a complex in general, as $s \in \Gamma(\mathcal{N}_{1,2})$ is not necessary a flat section.

In order to obtain a complex, we use the symplectic form to get an isomorphism $\lrcorner(i\omega_X) : \mathcal{N}_{1 \vee 2} \to T^*(B_{L_1} \cap B_{L_2})$ of bundles over $M_{1,2}$. This identification induces a map

$$\Omega_A^*((L_1, E_1), (L_2, E_2)) \to \Omega^*(M(L_1, L_2), \mathrm{Hom}(ev_s^* E_1, ev_t^* E_2)).$$

The operator D descends to give a cochain complex. The complex

$$(\Omega^*(M(L_1, L_2), \mathrm{Hom}(ev_s^* E_1, ev_t^* E_2)), D)$$

is defined to be the quantum intersection complex between two semi-flat A-branes.

Semi-flat B-branes

Similarly, we consider semi-flat complex submanifolds on $\check{X}$. A semi-flat complex submanifold $\check{L}$ is a submanifold of $\check{p}^{-1}(B_{\check{L}})$, where $B_{\check{L}}$ is a rational affine submanifold of B, such that $\check{p}|_{\check{L}} : \check{L} \to B_{\check{L}}$ is a torus fiber bundle over $B_{\check{L}}$, and each fiber over b is an affine subtorus of $\check{p}^{-1}(b)$.

Definition 4.6. A B-brane $(\check{L}, \check{E})$ on $\check{X}$ is called semi-flat if $\check{L}$ is a semi-flat complex submanifold and $\check{E}$ is a unitary line bundle on $\check{L}$, with its connection $\nabla_{\check{E}}$ satisfying $(\nabla_{\check{E}}|_{\check{p}^{-1}(b) \cap \check{L}})^2 = 0$ and $(\nabla_{\check{E}}^{0,1})^2 = 0$.

Given two semi-flat B-branes $(\check{L}_i, \check{E}_i)$, $i = 1, 2$, The space of fiberwise minimal geodesic paths, $\check{M}(\check{L}_1, \check{L}_2)$ is naturally a complex manifold. We define $\Omega_B^*((\check{L}_1, \check{E}_1), (\check{L}_2, \check{E}_2))$ using a similar construction. Again it is not a complex, due to the fact that $\check{E}_i$'s are not flat bundles. Notice that $\check{\mathcal{N}}_{1 \vee 2}$ is a holomorphic vector bundle over $\check{M}_{1,2}$, with a holomorphic section $\check{s}$ defining the operator $\check{\delta}$. Using the complex structure on $\check{M}_{1,2}$, we obtain a map given by projection to anti-holomorphic forms

$$\Omega_B^*((\check{L}_1, \check{E}_1), (\check{L}_2, \check{E}_2)) \to \Omega^{0,*}(\check{M}(\check{L}_1, \check{L}_2), \mathrm{Hom}(ev_s^* \check{E}_1, ev_t^* \check{E}_2) \otimes \wedge^* \check{\mathcal{N}}_{1 \vee 2}),$$

the operator $\check{D}$ descends to give the operator $\bar{\partial} + (-1)^{p+q+1}\check{\delta}$. The quantum intersection complex between two B-branes is given by

$$(\Omega^{0,*}(\check{M}(\check{L}_1, \check{L}_2), \mathrm{Hom}(ev_s^* \check{E}_1, ev_t^* \check{E}_2) \otimes \wedge^* \check{\mathcal{N}}_{1 \vee 2}), \check{D}).$$

4.4. Transform of semi-flat branes

We describe the transform between A- and B-branes. For simplicity, we transform an A-brane on X to give a B-brane on $\check{X}$. With the two projection maps

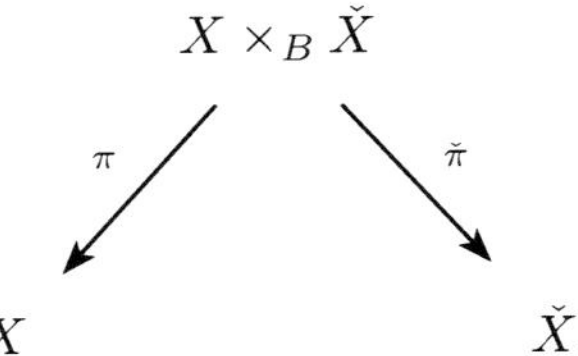

we denote the Poincaré line bundle on $X \times_B \check{X}$ by $\mathcal{P}$, which is constructed by combining fiberwise Poincaré line bundles into a family.

The Fourier transform is a family version of that given in section 3. Given a semi-flat A-brane (L, E), we define

$$\check{E}|_{\check{x}} = H^0\big(L \cap p^{-1}(\check{p}(\check{x})), \pi^* E \otimes \mathcal{P}^*|_{L \cap p^{-1}(\check{p}(\check{x}))}\big).$$

and

$$\check{L} = \{\check{x} \in \check{X} : \check{E}|_{\check{x}} \neq 0\}.$$

$\check{E}$ becomes a complex line bundle over $\check{L}$ and is equipped with a Hermitian metric and a unitary connection. The fact that E is a flat line bundle implies that $\check{L}$ is a semi-flat complex submanifold of $\check{X}$, and L being a Lagrangian submanifold implies that $\check{E}$ is a holomorphic line bundle over $\check{L}$. Furthermore, if L is a special Lagrangian submanifold, then $\check{E}$ is asymptotically a Hermitian Yang-Mills line bundle when $\check{X}$ is approaching a large complex structure limit. We define

$$\mathcal{F}(L, E) = (\check{L}, \check{E})$$

to be the mirror B-brane on $\check{X}$.

Transform of quantum intersection complexes

Given two semi-flat A-branes (L_i, E_i), $i = 1, 2$, with corresponding mirror B-branes $(\check{L}_i, \check{E}_i)$, we denote by $B_i \subset B$ the common base of the fibrations of L_i and $\check{L}_i$. We consider the fiber product

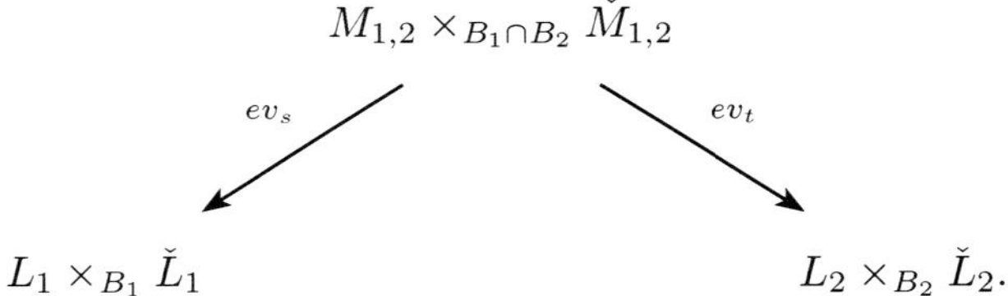

The kernel of Fourier transform $\phi_{\mathcal{F}} \otimes e^{G_1 + G_2}$ is defined by patching the fiberwise kernels into a relative version. The Fourier transform $\hat{\mathcal{F}}$ descends as a map

$$\Omega^*(M(L_1, L_2), \operatorname{Hom}(ev_s^* E_1, ev_t^* E_2))$$

$$\hat{\mathcal{F}} \downarrow$$

$$\Omega^{0,*}(\check{M}(\check{L}_1, \check{L}_2), \operatorname{Hom}(ev_s^* \check{E}_1, ev_t^* \check{E}_2) \otimes \wedge^* \check{\mathcal{N}}_{1\vee 2}).$$

The Fourier transform can be used to study the relation between the quantum intersection complexes of A- and B-branes.

5. Mirror symmetry and quantum corrections

The Lagrangian fibration $p : X \to B$ typically has singular fibers. The base of the fibration will be (possibly singular) integral affine manifold with corners, with Γ the critical locus (or discriminant locus) of p. Restricting to $B_0 = B - \Gamma$ and $X_0 = p^{-1}(B_0)$, we obtain a Lagrangian torus fiber bundle $p : X_0 \to B_0$. Roughly speaking, the appearance of singular fibers is due to the existence of vanishing cycles, and compactification data is captured by these vanishing cycles. These vanishing cycles are typically represented by holomorphic disks bounded by a loop in a fiber of $p|_{X_0}$.

In [8, 9], there are generating functions defined by "counting" holomorphic disks, or more precisely one-pointed open Gromov-Witten invariants [17, 18, 19], with boundary in fibers of p. These functions are used to construct the mirror manifold $\check{X}$. We give a brief review of these constructions below.

5.1. Mirror symmetry for toric Fano manifolds

Consider a n-dimensional toric Fano manifold X, with $P \subset M_{\mathbb{R}}$ its moment polytope given by

$$P = \bigcap_{j=1}^{d} \{b : (b, \nu_j) - \lambda_j \geq 0\},$$

and $\Gamma = \partial P$ the critical locus. We write

$$\mu : X \to P$$

for the moment map of X under the Hamiltonian T^n-action. The restriction of μ to the open dense orbit $X_0 \subset X$ is a torus bundle over the interior of the polytope P_0. This gives a Lagrangian torus fibration.

Kontsevich and Hori-Vafa [21] predicted that the mirror of a toric manifold X together with its symplectic structure ω_X is given by a Landau-Ginzburg model $(\check{X}, W)$, where $\check{X}$ is the non-compact Kähler manifold $(\mathbb{C}^*)^n$ and $W = \sum_{j=1}^{d} e^{\lambda_j - 2\pi(b + i\check{y}, \nu_j)}$ is a holomorphic function called the superpotential which is 'mirror' to information from the toric divisors at infinity. For example, $W = z_1 + z_2 + \frac{1}{z_1 z_2}$ is the superpotential for $\mathbb{C}P^2$.

In [9], the authors constructed a generating function $\Phi_q(x, v) \in C^\infty(X_0 \times N)$, by counting Maslov index 2 holomorphic disks bounded by the loop parametrized by (x, v) in $X_0 \times N$, where $X_0 \times N$ is interpreted as the space of fiberwise geodesic loops in X_0.

By taking $\check{X}$ to be the dual torus fibration of $\mu|_{X_0}$, fiberwise Fourier transform gives a function $\mathcal{F}(\Phi_q)$ on $\check{X} \times iM$. Since Φ_q is constant along the fibers of $\mu|_{X_0}$, $\mathcal{F}(\Phi_q)$ can be regarded as a function on $\check{X} \cong (\mathbb{C}^*)^n$. The symplectic structure on X can be incorporated to give the holomorphic structure on the Landau-Ginzburg model $(\check{X}, W)$ in the sense that

$$\mathcal{F}(\Phi_q e^{\omega_X}) = e^W \Omega_{\check{X}}.$$

Remark 5.1. The form $\Phi_q e^{\omega_X}$ can be viewed as the symplectic structure modified by quantum corrections from Maslov index 2 holomorphic disks in X with boundary on a Lagrangian torus fiber. The form $e^W \Omega_{\check{X}}$ should be viewed as the holomorphic volume form of the Landau-Ginzburg model $(\check{X}, W)$.

The isomorphism between the quantum cohomology of X and the Jacobian ring of $(\check{X}, W)$ can be established using the Fourier transform. For details, we refer readers to the article [9].

5.2. Mirror symmetry for toric Calabi-Yau manifolds

In [5, 8], a construction of the mirror manifold $\check{X}$ of toric Calabi-Yau manifolds (necessarily non-compact) is described, using the non-toric special Lagrangian fibrations constructed by M. Gross in [22]. The Gross fibration is introduced to investigate the appearance of singular fibers in the interior of the base B. It serves as local model to study a typical fibration of a Calabi-Yau manifold. For example, the Gross fibration for $\mathbb{C}^2$ will have as base B the upper half plane with boundary and a singular fiber over an interior point of B. The local structure for the singular fiber is a focus-focus singularity. We give a brief review for the construction of mirror manifolds given in [8].

The procedure involves constructing coordinate functions of $\check{X}$ by "counting" holomorphic disks emanating from boundary divisors of X. The problem is that in the Gross fibration, B has only one codimension-1 boundary. This is a technical issue and can be solved by modifying the Gross fibration using symplectic cuts. X appears as the limit as the extra divisors move to infinity.

26

The modified Gross fibration, which we continue to denote by $\mu : X \to B$, is a proper Lagrangian fibration whose base B is a polyhedral set in $M_{\mathbb{R}}$ having at least n distinct codimension-1 faces, which are denoted by Ψ_j for $j = 0, ..., m - 1$. The preimage $D_j := \mu^{-1}(\Psi_j)$ of each Ψ_j is assumed to be a codimension-2 submanifold in X. Furthermore, the critical locus besides Ψ_j is assumed to be contained in a hyperplane H. H is called 'the wall' and separates B_0 into two chambers.

Using the semi-flat Lagrangian fibration $\mu : X_0 \to B_0$, $\check{\mu} : \check{X}_0 \to B_0$ is defined by taking the dual torus fibration. The Lagrangian fibration μ defines a lattice bundle Λ over B_0, parametrizing fiberwise geodesic loops of $\mu : X_0 \to B_0$. For each j, a generating function $\mathcal{I}_j$ is defined on $\Lambda|_{B_0 - H}$ by counting holomorphic disks intersecting the submanifold D_j's, with boundary being the loop parametrized by $\Lambda|_{B_0 - H}$. The fiberwise Fourier transforms of $\mathcal{I}_j$'s give holomorphic functions $\check{z}^j$ on $\check{\mu}^{-1}(B_0 - H)$. $\check{z}^j$ changes dramatically from one component to another component, and this is called the wall-crossing phenomenon [5]. Let R be the subring of holomorphic functions on $\check{\mu}^{-1}(B_0 - H)$, generated by constant functions, $\check{z}^j$'s and $1/\check{z}^j$'s. The corrected mirror $\check{X}$ is defined to be $\check{X} = Spec(R)$.

For example, the mirror manifold $\check{X}$ for $K_{\mathbb{P}^2}$, is given by

$$\check{X} = \{(u, v, z_1, z_2) \in \mathbb{C}^2 \times (\mathbb{C}^*)^2 : uv = c(q) + z_1 + z_2 + \frac{q}{z_1 z_2}\},$$

where

$$c(q) = 1 - 2q + 5q^2 - 32q^3 + 286q^4 - 3038q^5 + \ldots,$$

with the coefficients defined by one-pointed open Gromov-Witten invariants.

Applying the above construction, the mirror manifold $\check{X}$ obtained belongs to the Hori-Iqbal-Vafa mirror family given in [23]. Moreover, a striking feature of the quantum-corrected mirror family constructed above is that it is inherently written in flat coordinates over the moduli space of complex structures of $\check{X}$ and gives an explicit description for the mirror map. For details, we refer readers to [8].

5.3. Semi-flat branes and quantum corrections

Finally, we mention an example where we can see how Fourier transform of semi-flat branes incorporates quantum corrections. We consider the case where X is a toric Fano manifold, equipped with moment map torus fibration. We use the same notations as in subsection 5.1.

For each $b \in P_0$, the fiber torus $\{b\} \times iT_N$ is a special Lagrangian submanifold. We consider a semi-flat A-brane $(\{b\} \times iT_N, E)$ where E is a flat unitary line bundle. The corresponding dual E-brane is a point $\check{z} = b + i\check{y} \in \check{X}$, with 1-dimensional vector space $\check{E} = H^0(\{b\} \times iT_N, E)$ over it. Then $M(\{b\} \times iT_N, \{b\} \times iT_N) \simeq T_N$ is the whole fiber torus and $\check{M}(\check{z}, \check{z}) \simeq M$ is the dual lattice of $\{b\} \times iT_N$. The quantum intersection complexes associated to $(\{b\} \times iT_N, E)$ and $(\check{z}, \check{E})$ are $\Omega^*(T_N)$ and $\Gamma(M, \wedge^* M_{\mathbb{C}})$, with differential operators given by d and δ respectively.

In [24], the authors showed that the Floer differential m_1 is given by $d + \iota$, where ι is defined by counting holomorphic disks of Maslov index 2. Given a Lagrangian torus fiber $\{b\} \times iT_N$, there are exactly d families of holomorphic disks of Maslov index 2 with boundary in $\{b\} \times iT_N$ (modulo to automorphisms of the domain). If we let $\{D_j\}_{j=1}^d$ be the holomorphic disks, then the operator ι is given by

$$\iota(\psi) = \sum_{j=1}^{d} [e^{-\int_{D_j} \omega_X} hol_{\nabla_E}^{-1}(\partial D_j)](\iota_{\partial D_j^\#} \psi),$$

where $\partial D_j^\#$ is the vector field generated by the circle action by $\partial D_j \in N$ and $hol_{\nabla_E}(\partial D_j)$ is the holonomy around the loop ∂D_j. The operator ι commutes with d.

Fourier transform gives an isomorphism

$$\hat{\mathcal{F}} : \Omega^*(T_M) \to \Gamma(M, \wedge^* M_{\mathbb{C}}).$$

If we identify $\wedge^* M_{\mathbb{C}} \simeq \wedge^* T_{\check{z}}^{1,0} \check{X}$ and use the result

$$\int_{D_j} \omega_X = 2\pi((b, \nu_j) - \lambda_j)$$

proven in [24], we deduce that $\hat{\mathcal{F}}(\iota)$ is precisely the operator given by contraction with ∂W, where W is the superpotential. This gives a geometric and direct verification of the result in [24].

6. Conclusion

We have seen that the *Fourier transform* plays an important role in various constructions in the SYZ programme, especially when dealing with quantum corrections, as the quantum corrections appear as "higher Fourier modes" for the mirror complex structures. To understand the construction of mirror manifolds and mirror correspondence of branes, it will be important to extend the definition of Fourier transform to flat branes, as quantum corrections for branes is also expected to incorporate with the Fourier transform. Furthermore, a geometric construction of the mirror correspondence of branes can help to explain certain mirror symmetric phenomena.

Acknowledgements: The second author thanks Akbulut for the invitation and being an excellent host for the Gökova/2011 workshop and the author benefited greatly from attending it. The authors thank Kwokwai Chan, Siu-Cheong Lau and Yi Zhang for useful discussions. The work described in this paper was substantially supported by a grant from the Research Grants Council of the Hong Kong Council of the Hong Kong Special Administrative Region, China (Project No. CUHK 401809).

References

[1] E. Witten, Mirror manifolds and topological field theory, *AMS/IP Studies in Advanced Mathematics* **9** (1998), 121–131.

[2] M. Kontsevich, Homological algebra of mirror symmetry, *Proceedings of the International Congress of Mathematicans* **1** (1994), 120–139.

[3] A. Strominger, S.-T. Yau, and E. Zaslow, Mirror symmetry is T-duality, *Nuclear Phys. B* **479** (1996), 243–259.

[4] D. Auroux, Mirror symmetry and T-duality in the complement of anticanonical divisor, *J. Gökova Geom. Topol.*, **1** (2007), 51–91.

[5] D. Auroux, Special Lagrangian fibrations, wall-crossing, and mirror symmetry. *Surveys in Differential Geometry*, **13** (2009), 1–47.

[6] B. Fang. Homological mirror symmetry is T-duality for $\mathbb{P}^n$, *Commun. Number Theory Phys*, **2** (2008), 719-742.

[7] B. Fang, C.-C. Liu, D. Treumann, and E. Zaslow, T-Duality and homological mirror symmetry of toric varieties, arXiv preprint, arXiv: 0811.1228.

[8] K.-W. Chan, S.-C. Lau and N. C. Leung, SYZ mirror symmetry for toric Calabi-Yau manifolds, *J. Differential Geom.*, to appear.

[9] K.-W. Chan and N. C. Leung, Mirror symmetry for toric Fano manifolds via SYZ transformations, *Adv. Math*, **223** (2010), 797–839.

[10] K.-W. Chan and N. C. Leung, On SYZ mirror transformation, *New Developments in Algebraic Geometry, Integrable Systems and Mirror Symmetry*, **59** (2010), 1–30.

[11] N. C. Leung, Mirror symmetry without corrections. *Comm. Anal. Geom.*, **13** (2005), 287–331.

[12] N. C. Leung, Geometric aspects of mirror symmetry (with SYZ for rigid CY manifolds), *Proceedings of ICCM*, 2001.

[13] N. C. Leung, S.-T. Yau and E. Zaslow, From special Lagrangian to Hermitian-Yang-Mills via Fourier-Mukai transform, *Adv. Theor. Math. Phys.*, **4** (2000), 1319-1341.

[14] A. Polishchuk and E. Zaslow, Categorical mirror symmetry: The elliptic curve, *Adv. Theor. Math. Phys.*, **2** (1998), 443-470.

[15] M. Gross, Toric degenerations and Batyrev-Borisov duality, *Mathematische Annalen*, **333** (2005), 645–688.

[16] M. Gross, The Strominger-Yau-Zaslow conjecture: From torus fibrations to degenerations, *Proceedings of Symposia in Pure Mathematics: Algebraic geometry, Seattle 2005*, **80** (2009), 149–192.

[17] K. Fukaya, Y.-G. Oh, H. Ohta, and K. Ono, Lagrangian Floer theory on compact toric manifolds I. *Duke Math. J.*, **151** (2010), 23–174.

[18] K. Fukaya, Y.-G. Oh, H. Ohta, and K. Ono, Lagrangian intersection Floer theory: anomaly and obstruction. Part I. *AMS/IP Studies in Advanced Mathematics*, **46**, 2009.

[19] K. Fukaya, Y.-G. Oh, H. Ohta, and K. Ono, Lagrangian intersection Floer theory: anomaly and obstruction. Part II. *AMS/IP Studies in Advanced Mathematics*, **46**, 2009.

[20] K.-L. Chan, N. C. Leung, and C. Ma, in preparation.

[21] K. Hori and C. Vafa, Mirror symmetry, arXiv preprint, arXiv:hep-th/0002222v3.

[22] M. Gross, Examples of special Lagrangian fibrations, in *Symplectic geometry and mirror symmetry*, World Sci. Publ., 2001, 81–109.

[23] K. Hori, A. Iqbal, and C. Vafa, D-branes and mirror symmetry, arXiv preprint, available at arXiv:hep-th/0005247.

[24] C.-H. Cho and Y.-G. Oh, Floer cohomology and disc instantons of Lagrangian torus fibers in Fano toric manifolds. *Asian J. Math.*, **10**(2006), 773–814.

THE INSTITUTE OF MATHEMATICAL SCIENCES AND DEPARTMENT OF MATHEMATICS, THE CHINESE UNIVERSITY OF HONG KONG, SHATIN, HONG KONG
E-mail address: leung@math.cuhk.edu.hk

THE INSTITUTE OF MATHEMATICAL SCIENCES AND DEPARTMENT OF MATHEMATICS, THE CHINESE UNIVERSITY OF HONG KONG, SHATIN, HONG KONG
E-mail address: klchan@math.cuhk.edu.hk

THE INSTITUTE OF MATHEMATICAL SCIENCES AND DEPARTMENT OF MATHEMATICS, THE CHINESE UNIVERSITY OF HONG KONG, SHATIN, HONG KONG
E-mail address: ncma@math.cuhk.edu.hk

Proceedings of 18th Gökova
Geometry-Topology Conference
pp. 31 – 41

Conway mutation and alternating links

Joshua Evan Greene

ABSTRACT. This paper is a conversational companion to *Lattices, graphs, and Conway mutation* [Gre11], on which I spoke at the 2011 Gökova Geometry/Topology Conference.

1. Introduction.

In his enumeration of small-crossing knots, Conway observed an interesting relationship between a particular pair of knot diagrams [Con70].

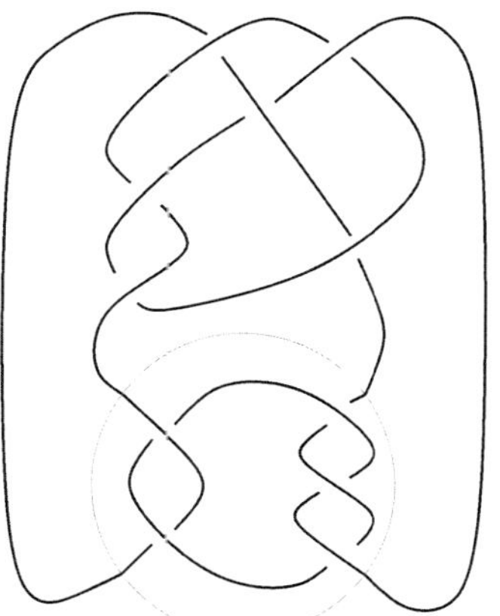 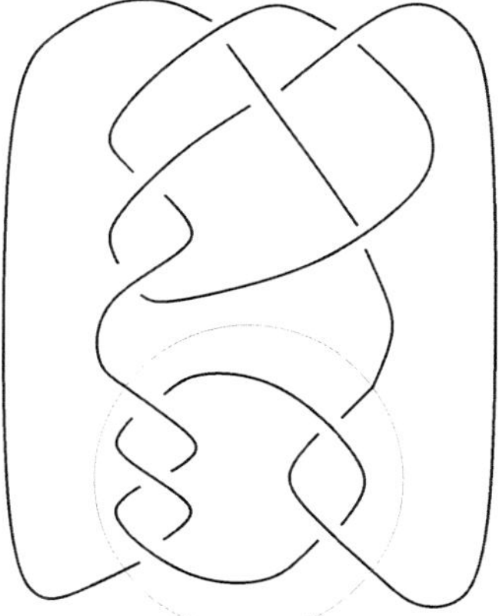

At left is the Kinoshita-Terasaka knot, and at right the one that now bears Conway's name. We get from one to the other by (a) excising the indicated disk, (b) rotating it by an angle π about a perpendicular axis piercing its center, and (c) reinserting it. Note that in place of (b) we could have instead rotated the disk about its horizontal or vertical axis to effect a transformation of diagrams; in the first case, the diagrams stay the same, and in the second case, they interchange. This operation is called an *elementary mutation*, and a pair of diagrams are called *mutants* if they are related by a sequence of isotopies and elementary mutations. In less diagrammatic terms, we have a sphere S^2 that meets a link $L \subset S^3$ transversely in four points, which we cut along and reglue by an involution

Key words and phrases. Knot theory, Heegaard Floer homology, lattices.

that fixes a pair of points disjoint from L and permutes $S^2 \cap L$. Doing so results in a new link $L' \subset S^3$, and a pair of *links* are called mutants if they are related by a sequence of isotopies and such transformations.

Mutant links are notoriously difficult to distinguish, since many invariants take the same values on them. Such invariants include the HOMFLY polynomial (and so the Alexander and Jones polynomials) and the knot signature (see [Lic97]). Other invariants do detect mutation, such as the Seifert genus, which Gabai calculated as 2 for the Kinoshita-Terasaka knot and 3 for the Conway knot [Gab86]. It is always an interesting problem to determine whether a given link invariant detects mutation.

The principal link invariant of interest in this paper is the homeomorphism type of the double-cover of S^3 branched along the link L. This space is called the *branched double cover* of L and is denoted by $\Sigma(L)$. This invariant falls into the category of those insensitive to mutation, as recorded in the following Proposition.

Proposition 1.1 (Viro [Vir76]). *If L and L' are mutant links, then $\Sigma(L) \cong \Sigma(L')$.*

Amusingly, Viro was unaware of Conway's work at the time he established this result!

Sketch of proof. It suffices to consider a pair of links L, L' related by mutation in a single Conway sphere S^2. Let τ denote the involution of S^2 relating L and L'. In both $\Sigma(L)$ and $\Sigma(L')$, the Conway sphere lifts to a torus T^2 which splits the space into two pieces whose boundaries get identified by an appropriate gluing map. The two pieces are the same for both $\Sigma(L)$ and $\Sigma(L')$, but the gluing maps differ by the lift of τ to the torus T^2. This lift is conjugate to a translation of T^2 (namely, one of the three translations of order two), so it is isotopic to the identity map. Hence the gluing maps are isotopic, and so $\Sigma(L) \cong \Sigma(L')$. $\square$

Thus, the homeomorphism type of $\Sigma(L)$ is more properly an invariant of the *mutation type of L*. From this perspective, we may ask how well $\Sigma(L)$ (or, indeed, any invariant of mutation type) distinguishes *between* mutation types. In general, $\Sigma(L)$ does not *completely* distinguish between mutation types. For example, the Brieskorn sphere $\Sigma(2, 3, 7)$ is the branched double cover of both the torus knot $T(3, 7)$ and the pretzel knot $P(-2, 3, 7)$. These knots are not mutants, since their respective Jones polynomials, $-q^{14} + q^8 + q^6$ and $-q^{13} + q^{12} - q^{11} + q^7 + q^5$, distinguish their mutation types. In fact, it remains a fascinating open problem to describe all pairs of links that have homeomorphic branched double covers. Many interesting constructions and partial results exist, yet no guiding conjecture has emerged.

However, for two-bridge links, $\Sigma(L)$ *is* a complete invariant of the mutation type and, moveover, the isotopy type. The essential reason is that an invariant of manifolds, the *Reidemeister-Franz torsion*, provides a complete invariant of the homeomorphism type of lens spaces. Indeed, two-bridge links and lens spaces are parametrized by pairs of relatively prime integers $p > q > 0$. The two-bridge link $S(p, q)$ has branched double cover $L(p, q)$, and there exists an isotopy $S(p, q) \simeq S(p, q')$ whenever $qq' \equiv 1 \pmod{p}$. Using the torsion invariant, one shows that $L(p, q) \cong L(p', q')$ iff $p = p'$ and either

$q = q'$ or $qq' \equiv 1 \pmod{p}$. This gives the stated result. Later, Bonahon and Hodgson-Rubinstein established the stronger result that a lens space admits a unique genus one Heegaard splitting up to isotopy. Furthermore, Hodgson-Rubinstein showed that if $\Sigma(L)$ is homeomorphic to a lens space, then L is isotopic to a two-bridge link.

Now, there is a mantram in knot theory which goes as follows.

Mantram 1.2. Generalize all questions and results about two-bridge links to alternating links.

Thus, in the present setting, we are led to ask whether $\Sigma(L)$ is a complete invariant of the mutation type of an alternating link, and whether the Reidemeister-Franz torsion distinguishes the homeomorphism types of these spaces. It turns out that there is a related invariant of 3-manifolds which is easier to manipulate for this broader class of manifolds: the *d-invariant in Heegaard Floer homology*. For the manifolds under consideration, the d-invariant recovers the torsion invariant.

Our main result states that, in fact, within the family of alternating links, the homeomorphism type of $\Sigma(L)$ is a complete invariant of the mutation type. Furthermore, the d-invariant is a complete invariant of the homeomorphism types of the spaces $\Sigma(L)$.

Theorem 1.3. *Given a pair of connected, reduced alternating diagrams D, D' for a pair of links L, L', the following assertions are equivalent:*

1. *D and D' are mutants;*
2. *L and L' are mutants;*
3. *$\Sigma(L) \cong \Sigma(L')$; and*
4. *the d-invariants of $\Sigma(L)$ and $\Sigma(L')$ are the same.*

(The equivalence (1) $\Longleftrightarrow$ (2) is originally due to Menasco [Men84].)

One consequence of Theorem 1.3 is that an alternating link without an essential Conway sphere has a unique reduced, alternating diagram up to isotopy. This, of course, follows from the Menasco-Thistlethwaite theorem (formerly, the Tait flyping conjecture), but our argument in this case is substantially different. In fact, Theorem 1.3 motivates the following question.

Question 1.4. Is there a natural Floer-theoretic invariant that distinguishes isotopy types of alternating links?

We emphasize that Theorem 1.3 says *nothing* about the case of a non-alternating link. In fact, we make the following conjecture.

Conjecture 1.5. *If a pair of links have homeomorphic branched double-covers, then either both are alternating or both are non-alternating.*

The Hodgson-Rubinstein result implies the validity of Conjecture 1.5 in the case that one of the links is a two-bridge link. Besides that result, our strongest evidence in support of Conjecture 1.5 is the lack of any counterexample! If true, its proof would undoubtedly require techniques outside the scope of Heegaard Floer homology. Indeed, there exist infinitely many examples of prime links L, L' where L is alternating and L' is non-alternating, yet $\Sigma(L)$ and $\Sigma(L')$ have identical Floer invariants.

In the next section, we describe the d-invariant and how it is calculated for the manifolds of interest. We close this Introduction by remarking that we do not know how to establish the equivalence (2) $\iff$ (3) of Theorem 1.3 without the use of the d-invariant. Is there some more topological argument, a là Bonahon and Hodgson-Rubinstein, that could establish it, and perhaps resolve Conjecture 1.5 as well?

2. The d-invariant.

2.1. The input from Floer theory.

All the major definitions and results in this section are due to Ozsváth-Szabó. Throughout, let Y denote a closed, oriented, rational homology sphere. The d-invariant of Y is an invariant derived from the Heegaard Floer homology of Y. Since its introduction, it has played an instrumental role in many applications of Heegaard Floer homology to low-dimensional topology, including problems pertaining to knot concordance, Dehn surgery, the unknotting number, and, as we have here, a classification result.

In short, it is a mapping $d : \mathrm{Spin}^c(Y) \to \mathbb{Q}$, where $\mathrm{Spin}^c(Y)$ denotes the set of spin^c structures on Y. The set of spin^c structures on a 3- or 4-manifold forms a torsor over the integral second cohomology group, and we shall see that they are quite easy to manipulate, if a little tricky to define (as we shall avoid doing). Spin^c structures occur naturally in Heegaard Floer homology. The hat version of this theory assigns to a pair $(Y, \mathfrak{t})$, $\mathfrak{t} \in \mathrm{Spin}^c(Y)$, a finite-dimensional vector space over $\mathbb{Z}/2\mathbb{Z}$ denoted $\widehat{HF}(Y, \mathfrak{t})$. This group is graded by rational numbers, and the invariant $d(Y, \mathfrak{t}) \in \mathbb{Q}$ records a distinguished grading on this group.

For us, the most important property of the d-invariant is its relationship with the four-dimensional theory. Suppose that X is a smooth manifold with boundary Y and a positive[1] definite intersection pairing. Suppose, furthermore, that $H_1(X; \mathbb{Z}) = 0$, so that every spin^c structure on Y extends to one on X (by analogy to the case for cohomology). Then, for all $\mathfrak{t} \in \mathrm{Spin}^c(Y)$, we have an inequality

$$d(Y, \mathfrak{t}) \leq \min \left\{ \frac{c_1(\mathfrak{s})^2 - b_2(X)}{4} \,\middle|\, \mathfrak{s} \in \mathrm{Spin}^c(X), \, \mathfrak{s}|Y = \mathfrak{t} \right\}, \tag{1}$$

where we make use of the *first Chern class* mapping $c_1 : \mathrm{Spin}^c(X) \to H^2(X; \mathbb{Z})$. The group $H^2(X; \mathbb{Z})$ comes equipped with a symmetric, bilinear, $\mathbb{Q}$-valued pairing induced by the cup product, thereby making sense out of the term $c_1(\mathfrak{s})^2$ (much more on this in Subsection 2.3).

We call the manifold X *sharp* when inequality (1) is an equality for every $\mathfrak{t} \in \mathrm{Spin}^c(Y)$. Sharp 4-manifolds are a prized possession of anyone attempting to calculate the d-invariant. The d-invariant of a space Y is typically quite complicated to compute, but when Y is

---

[1] Practitioners of Floer homology typically orient their definite manifolds to be *negative* definite, but as we shall focus on the resulting algebraic/combinatorial story, we prefer positive definite ones.

34

known to bound a sharp 4-manifold X, we can extract it in terms that depend solely on the intersection pairing on $H_2(X; \mathbb{Z})$.

2.2. Branched double covers and graph lattices.

As luck would have it, the branched double cover of a non-split alternating link L bounds a very natural sharp 4-manifold. Begin with an alternating diagram D of L. The diagram splits the plane into regions, which we may color black and white in checkerboard fashion (see Figure 1 in page 38). The union of the black regions, together with a half-twist at each crossing, produces a (typically non-orientable) spanning surface F with boundary L. Place the link $L \subset S^3$ in the boundary of a 4-ball D^4, and push the interior of F slightly into the interior of D^4. Then the double cover of D^4, branched along the pushed-in copy of F, is a sharp 4-manifold X_D with boundary $\Sigma(L)$. (We are cheating a little bit on orientations here, but never mind).

The intersection pairing on $H_2(X_D)$ has a very nice description in terms of the combinatorics of D. Construct a planar graph G by placing a vertex in each white region and an edge between a pair of vertices whenever their regions touch at a crossing (Figure 2). This graph is called the *Tait graph* of the diagram D, and the intersection pairing on $H_2(X_D)$ is captured by the *lattice of flows* on the Tait graph, $\mathcal{F}(G)$.

Let us elaborate on what this means. Given a graph G, we orient its vertices and edges arbitrarily so as to give it the structure of a 1-dimensional CW complex. Letting $e_1, \ldots, e_m$ denote its oriented edges, we give the chain group $C_1(G; \mathbb{Z})$ the structure of a lattice by declaring that these edges form an orthonormal basis for it. Now the cycle space $Z_1(G; \mathbb{Z}) = \ker(\partial_1 : C_1(G; \mathbb{Z}) \to C_0(G; \mathbb{Z}))$ inherits the structure of a sublattice of $C_1(G; \mathbb{Z})$, and its isomorphism type is the lattice of flows $\mathcal{F}(G)$. We point out that an alternate choice of orientation on the vertices and edges of G does not affect the isomorphism type of $\mathcal{F}(G)$.

The lattice of flows on a graph has a natural companion called the *lattice of cuts* $\mathcal{C}(G)$. This is the sublattice of $C_1(G; \mathbb{Z})$ arising as the image of the adjoint mapping $\partial_1^* : C_0(G; \mathbb{Z}) \to C_1(G; \mathbb{Z})$. Together, $\mathcal{F}(G)$ and $\mathcal{C}(G)$ form a pair of complementary sublattices of $C_1(G; \mathbb{Z})$ that interact in a very nice way that we utilize in Subsection 3.2.

2.3. Lattices and intersection pairings.

Returning to our earlier level of generality, let X denote a smooth, compact 4-manifold with $H_1(X; \mathbb{Z}) = 0$ and boundary a rational homology sphere Y. The intersection pairing on homology gives the group $H_2(X; \mathbb{Z})$ the structure of an *integral lattice*, that is, a free abelian group equipped with a symmetric, bilinear, non-degenerate, integer-valued pairing. A portion of the long exact sequence in integral cohomology reads

$$0 \to H_2(X) \to H^2(X) \to H^2(Y) \to 0.$$

Here we have substituted the term $H_2(X)$ for $H^2(X, \partial X)$ using Poincaré duality. The Hom-pairing endows the group $H^2(X)$ with the structure of the *dual* lattice to $H_2(X)$, so endowing it with a symmetric, bilinear, *rational*-valued pairing. The quotient of $H^2(X)$

by $H_2(X)$ is isomorphic to the finite, abelian group $H^2(Y)$. In purely lattice-theoretic terms, the set-up is modeled by the sequence

$$0 \to \Lambda \to \Lambda^* \to \overline{\Lambda} \to 0,$$

where Λ denotes an integral lattice, Λ^* the dual lattice $\{\lambda^* \in \Lambda \otimes \mathbb{Q} \mid \langle \lambda^*, \lambda \rangle \in \mathbb{Z}, \, \forall \lambda \in \Lambda\}$, and $\overline{\Lambda}$ the *discriminant group* (or *dual quotient group*) Λ^*/Λ.

To pull spin^c structures into the picture, we make two constructions. The dual lattice Λ^* contains a distinguished subset

$$\mathrm{Char}(\Lambda) = \{\chi \in \Lambda^* \mid \langle \chi, \lambda \rangle \equiv |\lambda| \,(\mathrm{mod}\ 2), \, \forall \lambda \in \Lambda\},$$

the set of *characteristic elements* for Λ. In terms of a fixed basis $\{\lambda_1, \ldots, \lambda_n\}$ for Λ, this set is expressed with respect to the dual basis as $\{\sum \chi_i \lambda_i^* \in \Lambda^* \mid \chi_i \equiv |\lambda_i| \,(\mathrm{mod}\ 2), \, \forall i\}$. We see that $\mathrm{Char}(\Lambda)$ is a coset of $2\Lambda^*$ in Λ^*, so comes with a natural free, transitive action by Λ^*: an element λ^* shifts χ by *twice* λ^*. In other words, $\mathrm{Char}(\Lambda)$ forms a torsor over Λ^*. Lastly, we define

$$C(\Lambda) = \mathrm{Char}(\Lambda) \,(\mathrm{mod}\ 2\Lambda).$$

This set forms a torsor over $\overline{\Lambda}$.

The significance of these constructions in the present setting is as follows. The first Chern class mapping sets up a one-to-one correspondence

$$c_1 : \mathrm{Spin}^c(X) \overset{\sim}{\to} \mathrm{Char}(\Lambda) \subset \Lambda^*,$$

where $\Lambda = H_2(X)$. Furthermore, there is a correspondence $C(\Lambda) \to H^2(Y)$, under which the restriction mapping $\mathrm{Spin}^c(X) \to \mathrm{Spin}^c(Y)$ is the reduction $\mathrm{Char}(\Lambda) \to C(\Lambda)$. Furthermore, the torsor structure $\mathrm{Spin}^c(X)/H^2(X)$ is modeled by $\mathrm{Char}(\Lambda)/\Lambda^*$, and $\mathrm{Spin}^c(Y)/H^2(Y)$ is modeled by $C(\Lambda)/\overline{\Lambda}$.

Example. Let X_k denote the D^2-bundle over S^2 with Euler number $k \neq 0$. Thus, $H_1(X_k; \mathbb{Z}) = 0$, and $\partial X_k \cong -L(k, 1)$, a lens space. Write Λ_k for the lattice $H_2(X_k)$, so Λ_k is generated by a single element λ with norm k. It follows that Λ^* is generated by a single element λ^* such that $\langle \lambda^*, \lambda \rangle = 1$. This element has norm $1/k$. We have $\mathrm{Char}(\Lambda_k) = \{j\lambda^* \mid j \equiv k \,(\mathrm{mod}\ 2)\}$. The classes in $C(\Lambda)$ take the form $(j + 2k\mathbb{Z})\lambda^*$ for the k different values $j \,(\mathrm{mod}\ 2k)$, $j \equiv k \,(\mathrm{mod}\ 2)$. This set forms a torsor over the discriminant group $\overline{\Lambda_k} \cong \mathbb{Z}/k\mathbb{Z}$.

Now suppose that the pairing on Λ is positive definite. Given a class $[\chi] \in C(\Lambda)$, we define

$$d_\Lambda([\chi]) = \min\left\{ \left. \frac{|\chi'| - \mathrm{rk}(\Lambda)}{4} \,\right|\, \chi' \in [\chi] \right\}.$$

Thus, we obtain a mapping $d_\Lambda : C(\Lambda) \to \mathbb{Q}$, the *d-invariant of the lattice* Λ. This definition is obviously modeled on the d-invariant in Heegaard Floer homology. Let Λ be $H_2(X)$, then the correspondence $C(\Lambda) \overset{\sim}{\to} \mathrm{Spin}^c(Y)$, $[\chi] \mapsto \mathfrak{t}$ gives us $d_\Lambda([\chi]) \leq d(Y, \mathfrak{t})$. Furthermore, if X is sharp, then $d_\Lambda([\chi]) = d(Y, \mathfrak{t})$, for all $\mathfrak{t} \in \mathrm{Spin}^c(Y)$. Based on this definition, we call an element $\chi \in \mathrm{Char}(\Lambda)$ *short* if its norm is minimal in $[\chi]$, and denote the set of short elements by $\mathrm{Short}(\Lambda)$.

Example (continued) We are going to calculate the d-invariant of the lens spaces $-L(k, 1)$, $k \geq 1$. When $k \geq 1$ the lattice Λ_k is positive definite. Each class in $C(\Lambda_k)$ contains a unique representative $j\lambda^*$ with $|j| \leq k$, except for the class containing $\pm k\lambda^*$, which contains these two. Altogether, these $k + 1$ elements constitute $\mathrm{Short}(\Lambda_k)$, and it follows that the d-invariant of Λ takes the values $(j^2/k - 1)/4$ on these respective classes. The manifold X_k is sharp for all $k \geq 1$, so under the correspondence between $\mathrm{Spin}^c(-L(k, 1))$ and $\{-k < j \leq k, \, j \equiv k \, (\mathrm{mod}\ 2)\}$, we have $d(-L(k, 1), j) = (j^2/k - 1)/4$.

We conclude this subsection on lattices with a very important fact. Note that for the case of the lattice $\mathbb{Z} = \Lambda_1$, the d-invariant takes the value 0 on the unique class in $C(\mathbb{Z})$. This is the case as well for the lattice $\mathbb{Z}^n$, which has an orthonormal basis $\{e_1, \ldots, e_n\}$: in this case, $\mathrm{Char}(\mathbb{Z}^n) = \{\sum \chi_i e_i \mid \chi_i \equiv 1 \, (\mathrm{mod}\ 2), \, \forall i\}$, and $\mathrm{Short}(\mathbb{Z}^n)$ consists of $\{\sum \chi_i e_i \mid |\chi_i| = 1, \, \forall i\}$, so indeed $d_{\mathbb{Z}^n}([\chi]) = (n - n)/4 = 0$. A powerful result due to Elkies asserts the converse to this observation for the case of a unimodular lattice Λ (i.e., $\Lambda = \Lambda^*$).

Theorem 2.1 (Elkies [Elk95]). *If Λ is a rank n, unimodular, integral, positive definite lattice with d-invariant 0, then $\Lambda \cong \mathbb{Z}^n$, i.e., Λ admits an orthonormal basis.* $\qquad \Box$

3. Diagrams, graphs, and lattices.

The groundwork laid over the course of the preceding section can be summarized in the following result.

Theorem 3.1 (Ozsváth-Szabó [OSz05], Thm. 3.4). *Let L denote a non-split alternating link and G the Tait graph of an alternating diagram of L. Then the d-invariant of $\Sigma(L)$ is calculated as minus the d-invariant of $\mathcal{F}(G)$.* $\qquad \Box$

Where does this result put us? We would like to show that the d-invariant of $\Sigma(L)$ determines the mutation type of a reduced, alternating diagram D for L. In light of Theorem 3.1, our task now is to first recast mutation of alternating diagrams in terms of some equivalence relation on graphs, and then to characterize the equivalence classes under this relation in terms of the d-invariant on the flow lattice.

3.1. Mutation and 2-isomorphism.

Let us consider the effect that mutation has on the Tait graph. Figure 1 displays four different link diagrams, each pair of which are related by an appropriate mutation in the indicated disk. Figure 2 displays their Tait graphs, which are distinctly embedded planar graphs. However, both the top two and the bottom two diagrams give rise to isomorphic Tait graphs as *abstract* graphs. We say that their two planar embeddings differ by a *flip*. This happens when a circle in the plane meets the planar graph G in a two-vertex cutset, and we simply reflect the embedding of G interior to the disk it bounds. The Tait graphs top and bottom are actually distinct from one another. They are related by a *planar switch*. This happens when a circle in the plane meets a planar graph G in a two-vertex cutset, and we rotate its interior by $180°$ (possibly composing with a reflection).

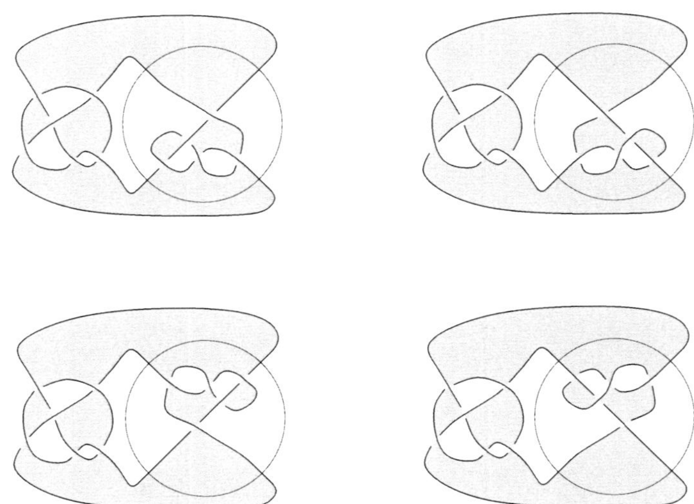

FIGURE 1. Four alternating link diagrams related by mutation.

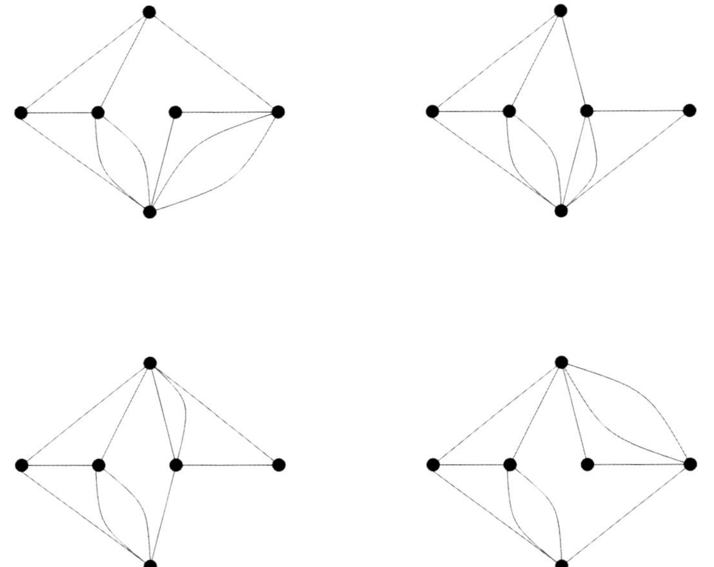

FIGURE 2. Four Tait graphs related by flips and switches.

Thus, mutant diagrams give rise to embedded Tait graphs that are related by a sequence of flips and planar switches.

At the level of abstract graphs, a planar switch is an instance of an *abstract switch*. This happens when we have a pair of graphs H_1, H_2, each with a pair of distinguished vertices $v_i, w_i \in V(H_i)$, $i = 1, 2$. We can glue together the vertices between the pairs $\{v_1, w_1\}$ and $\{v_2, w_2\}$ in either of two ways to form two graphs G, G', and these two are said to differ by a switch. A switch between a pair of graphs G and G' gives rise to a bijection between the edge sets $E(G) \overset{\sim}{\to} E(G')$. This bijection is far from arbitrary: it preserves edge sets of cycles, as one can check in Figure 2. A cycle-preserving bijection between the edge sets of a pair of graphs is called a *2-isomorphism*.

Now, a fundamental result due to Whitney asserts that a 2-isomorphism between a pair of abstract graphs is a composition of switches. Another fundamental result (essentially) due to Whitney asserts that any two planar drawings of a planar graph are related by a sequence of flips. Collecting these facts with the preceding observations, we have the following graph-theoretic characterization of mutation type.

Theorem 3.2 (Whitney $+ \epsilon$ [Whi32, Whi33]). *The Tait graph construction establishes a 1-1 correspondence*

$$\left\{ \frac{\text{alternating diagrams}}{\text{mutation}} \right\} \overset{\sim}{\to} \left\{ \frac{\text{planar graphs}}{\text{2-isomorphism}} \right\}.$$

(More precisely, the alternating diagrams should be connected and reduced, and the planar graphs should be 2-connected.)

3.2. 2-isomorphism and the d-invariant.

In view of Theorem 3.2, we now want to say that the d-invariant of the lattice $\mathcal{F}(G)$ determines the 2-isomorphism type of G. This is the content of the following result.

Theorem 3.3. *The following are equivalent for a pair of 2-edge-connected graphs G, G':*

    (1) *G and G' are 2-isomorphic;*
    (2) *$\mathcal{F}(G) \cong \mathcal{F}(G')$; and*
    (3) *$\mathcal{F}(G)$ and $\mathcal{F}(G')$ have isomorphic d-invariants.*

Equipped with Theorem 3.3, we can quickly wrap up the proof of Theorem 1.3 as follows. Given a pair of reduced, alternating diagrams D, D' representing links L, L' such that $\Sigma(L)$ and $\Sigma(L')$ have the same d-invariant, the corresponding Tait graphs G, G' are 2-isomorphic by Theorems 3.1 and 3.3, and so we conclude that the diagrams D, D' are mutants by Theorem 3.2.

Thus, Theorem 3.3 is the combinatorial heart of the matter. There are three principal ingredients that go into its proof: the *lattice gluing* construction, Elkies's Theorem 2.1, and the complementary nature of the lattices $\mathcal{F}(G)$ and $\mathcal{C}(G)$.

The instance of lattice gluing that we utilize goes as follows. Suppose that we have an isomorphism of discriminant groups $\varphi : \overline{\Lambda_1} \overset{\sim}{\to} \overline{\Lambda_2}$, where Λ_1, Λ_2 denote a pair of integral lattices. Then we can form a new lattice

$$\Lambda = \Lambda_1 \oplus_\varphi \Lambda_2 := \{\lambda_1^* + \lambda_2^* \in \Lambda_1^* \oplus \Lambda_2^* \mid \varphi(\overline{\lambda_1^*}) + \overline{\lambda_2^*} = 0\}.$$

The glue lattice Λ is integral and unimodular, it contains $\Lambda_1 \oplus \Lambda_2$ as a sublattice, and it will be definite provided that Λ_1 and Λ_2 are.

If we ask for more than just an isomorphism φ, then we get a corresponding payoff. Thus, suppose that Λ_1 and Λ_2 are definite lattices, and we have an isomorphism of torsors $C(\Lambda_1) \xrightarrow{\sim} C(\Lambda_2)$ that carries d_{Λ_1} into $-d_{\Lambda_2}$. Then we can carry out the above construction, and in this case the glue lattice Λ that results will have a vanishing d-invariant: hence $\Lambda \cong \mathbb{Z}^n$, thanks to Elkies's Theorem! Furthermore, it turns out that every $\chi_1 \in \mathrm{Short}(\Lambda_1)$ adds with some $\chi_2 \in \mathrm{Short}(\Lambda_2)$ to produce $\chi_1 + \chi_2 \in \mathrm{Short}(\mathbb{Z}^n)$.

The setup of the preceding paragraph applies, in particular, to the pair of lattices $\mathcal{F}(G)$ and $\mathcal{C}(G)$. (This is the nice interaction between these two to which we alluded at the end of Subsection 2.2.) But gluing them together just results in the lattice $C_1(G; \mathbb{Z})$ with $\mathcal{F}(G) \oplus \mathcal{C}(G)$ embedded standardly within it, so we have not really learned anything.

But now let us suppose that G and G' are graphs for which $\mathcal{F}(G)$ and $\mathcal{F}(G')$ have the same d-invariant. Then it follows that the preceding setup applies to the pair of lattices $\mathcal{F}(G)$ and $\mathcal{C}(G')$, and we can apply the gluing construction to *this* pair of lattices and actually get somewhere. Thus, we get an embedding $\mathcal{F}(G) \oplus \mathcal{C}(G') \hookrightarrow \mathbb{Z}^n$ with the property every element in $\mathrm{Short}(\mathcal{F}(G))$ adds with some element in $\mathrm{Short}(\mathcal{C}(G'))$ to give an element in $\mathrm{Short}(\mathbb{Z}^n)$. A subtle combinatorial argument then implies that the embedding $\mathcal{F}(G) \hookrightarrow \mathbb{Z}^n$ factors as the composite $\mathcal{F}(G) \hookrightarrow C_1(G; \mathbb{Z}) \xrightarrow{\sim} \mathbb{Z}^n$ for a suitable isomorphism between the latter, and similarly for $\mathcal{C}(G')$ [Gre11, Prop.2.8]. Now the isomorphisms $C_1(G; \mathbb{Z}) \xrightarrow{\sim} \mathbb{Z}^n \xleftarrow{\sim} C_1(G'; \mathbb{Z})$ set up a bijection between the edge sets of G and G'. As one might guess, this bijection establishes the desired 2-isomorphism, and arguing that it is a 2-isomorphism is actually quite easy!

This sketch establishes the implication $(3) \implies (1)$ of Theorem 3.3, which suffices for the desired application to Theorem 1.3. We point out that the implication $(2) \implies (3)$ is immediate, and that $(1) \implies (2)$ is a nice exercise (see [BdlHN97, Prop.5]).

4. Conclusion.

The proof sketch we have given of Theorem 1.3 seems tailor-made to the case of alternating links. Is there any natural, broader class of links to which it applies? It seems essential that the spaces $\Sigma(L)$ bound a sharp 4-manifold with both orientations. This motivates a question.

Question 4.1. Suppose that Y is a rational homology sphere, and Y bounds a sharp 4-manifold with both orientations. Does it follow that $Y \cong \Sigma(L)$ for some alternating link L?

If this were the case, and in addition Conjecture 1.5 were true, then we would obtain a non-diagrammatic characterization of alternating links, albeit in very round-about terms.

Acknowledgements: Thanks to the organizers of and participants in the 2011 Gökova Geometry/Topology Conference for a terrific week of mathematics and diving, and especially to Selman Akbulut for encouraging me to write these notes.

References

[BdlHN97] R. Bacher, P. de la Harpe, and T. Nagnibeda, The lattice of integral flows and the lattice of integral cuts on a finite graph, *Bull. Soc. Math. France*, **125** (1997), no. 2, 167–198.

[Con70] J. H. Conway, An enumeration of knots and links, and some of their algebraic properties, in Computational Problems in Abstract Algebra (Proc. Conf., Oxford, 1967), Pergamon, Oxford, 1970, 329–358.

[Elk95] N. D. Elkies, A characterization of the $\mathbb{Z}^n$ lattice, *Math. Res. Lett.*, **2** (1995), no. 3, 321–326.

[Gab86] D. Gabai, Genera of the arborescent links, *Mem. Amer. Math. Soc.*, no. 339, 1986, i–viii and 1–98.

[Gre11] J. E. Greene Lattices, graphs, and Conway mutation, arXiv preprint, arXiv:1103.0487.

[Lic97] W. B. R. Lickorish, *An introduction to knot theory*, Graduate Texts in Mathematics, vol. 175, Springer-Verlag, New York, 1997.

[Men84] W. Menasco, Closed incompressible surfaces in alternating knot and link complements, *Topology*, **23** (1984), no. 1, 37–44.

[OSz05] P. Ozsváth and Z. Szabó, On the Heegaard Floer homology of branched double-covers, *Adv. Math.*, **194** (2005), no. 1, 1–33.

[Vir76] O. Ja. Viro, Nonprojecting isotopies and knots with homeomorphic coverings, *Zap. Naučn. Sem. Leningrad. Otdel. Mat. Inst.*, **66** (1976), English translation in *J. Soviet Math.*, **12** (1979), 86–96.

[Whi32] H. Whitney, Congruent graphs and the connectivity of graphs, *Amer. J. Math.*, **54** (1932), no. 1, 150–168.

[Whi33] ———, 2-isomorphic graphs, *Amer. J. Math.*, **55** (1933), 245–254.

DEPARTMENT OF MATHEMATICS, BOSTON COLLEGE, CHESTNUT HILL, MA 02467
E-mail address: joshua.greene@bc.edu

Action of the cork twist on Floer homology

Selman Akbulut and Çağrı Karakurt

ABSTRACT. We utilize the Ozsváth-Szabó contact invariant to detect the action of involutions on certain homology spheres that are surgeries on symmetric links, generalizing a previous result of Akbulut and Durusoy. Potentially this may be useful to detect different smooth structures on 4-manifolds obtained by the cork twisting operation.

1. Introduction

Any two different smooth structures of a closed simply connected 4-manifold are related to each other by a cork twisting operation [16], and the cork can be assumed to be a Stein manifold [4] (see [9] and [10] for applications). A quick way to generate corks, which was used in [11], is from symmetric links as follows: Let L be a link in S^3 with two components $K_1 \cup K_2$. Suppose that L satisfies the following:

(**1**) Both components K_1 and K_2 are unknotted.
(**2**) There is an involution of S^3 exchanging K_1 and K_2.
(**3**) The linking number of K_1 and K_2 is ± 1 (for some choice of orientations).

From this we can construct a 4–manifold $W(L)$ by carving out a disk bounded by K_1 from a 4–ball, and attaching a 2–handle along K_2 with framing 0. Therefore a handlebody diagram of $W(L)$ is given by a planar projection of L decorated by a dot on K_1 (cf. [1]), and a 0 on top of K_2. We also require that the 4–manifold $W(L)$ admits an additional structure:

(**4**) The handlebody of $W(L)$ described above is induced by a Stein structure.

The last condition can be reformulated as follows:

(**4'**) Regard $K_2 \subset S^1 \times S^2 = \partial S^1 \times B^3$ equipped with the unique Stein fillable contact structure. Then the maximal Thurston-Bennequin number of K_2 is at least $+1$.

1991 *Mathematics Subject Classification.* 58D27, 58A05, 57R65.
The first named author is partially supported by NSF grant DMS 9971440.
The second named author is supported by a Simons fellowship.
Both authors are partially supported by NSF Focused Research Grant DMS-1065955.

We will call links satisfying conditions $(1) - (4)$ *admissible.* Some examples of admissible links are given in Figure 1. These examples were first studied in [9]. Note that the Hopf link is not admissible as it does not satisfy condition **(4)** even though the corresponding 4–manifold is B^4 which admits a Stein structure. Condition **(3)** ensures that $W(L)$ a contractible 4–manifold. Hence its boundary is a homology sphere. By condition **(2)**, we have an involution τ on $\partial W(L)$ that is obtained by exchanging the components of L. The significance this involution is indicated by the following theorem:

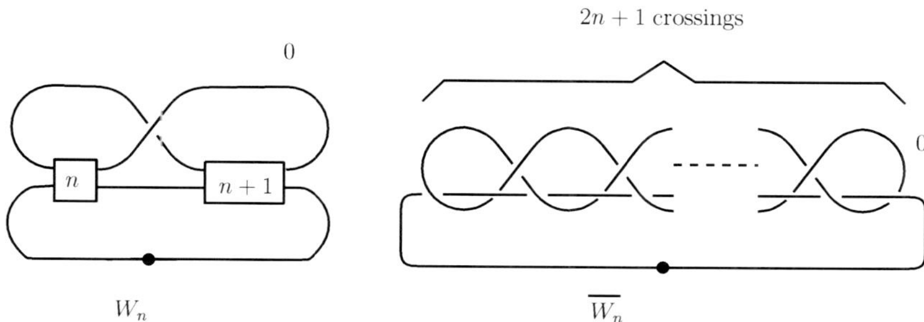

FIGURE 1. Examples of corks associated to certain symmetric links

Theorem 1.1. *Let L be an admissible link. The involution $\tau : \partial W(L) \to \partial W(L)$ acts non trivially on the Heegaard Floer homology group $HF^+(-\partial W(L))$*

This result generalizes [6], and it implies a result from [11]; namely τ does not extend to $W(L)$ as a diffeomorphism, even though it extends as a homeomorphism. Therefore $W(L)$ is a cork in the sense of [9]. The involution τ is called the *cork twist.* Theorem 1.1 was proved for the Mazur manifold W_1 in [2], and [26] (for instanton Floer homology), and [6] (for Heegaard Floer homology). Unlike the arguments in these papers, we do not explicitly find the Floer homology of $\partial W(L)$, and calculate the homomorphism the cork twist induces. Instead, we show that the homomorphism permutes two different distinguished Floer homology classes. These classes $c^+(\xi)$ are naturally associated to a cork via the induced contact structure ξ on the boundary. This suggests that a cork should be considered with its contact homology class $(W, c^+(\xi))$, to be used as a tool for checking nontriviality of its involution. In the course of our proof, we will incorporate several techniques developed in [9], [8] and [25].

Organization of this paper is as follows. We will review some standard facts about Stein manifolds, and Heegaard Floer homology respectively in the next two sections. We shall deduce Theorem 1.1 from a slightly stronger result (Theorem 4.1). This result along with some easy consequences are discussed in section 4.

2. Stein manifolds and their symplectic compactifications

Our aim in this section is to review the proof of the embedding theorem of Stein manifolds into closed symplectic manifolds in dimension four as given in [8]. See [18] for an alternative proof. We assume that the reader is familiar with the basics of contact geometry, open book decompositions and Lefschetz fibrations (cf. [19]). We start by recalling the topological characterization of Stein manifolds.

Theorem 2.1 ([12]). *Let* $W = B^4 \cup (1\text{-}handles) \cup (2\text{-}handles)$ *be a* 4-*dimensional handlebody with one* 0-*handle and no* 3 *or* 4-*handles. Then:*

(1) *The standard Stein structure on* B^4 *can be extended over the* 1-*handles so that the manifold* $W_1 := B^4 \cup (1\text{-}handles)$ *becomes Stein.*

(2) *If each* 2-*handle is attached to* ∂W_1 *along a Legendrian knot with framing one less than the Thurston-Bennequin framing, then the complex structure on* W_1 *can be extended over the* 2-*handles making* W *a Stein manifold.*

(3) *The handle decomposition of* W *is induced by a strictly plurisubharmonic Morse function.*

By Theorem 2.1 (see also [15]), we represent Stein manifolds by special kind of handlebody diagrams which contain handles of index up to two and the attaching circles of the two handles are all Legendrian (i.e., they have horizontal cusps instead of vertical tangencies and the smaller slope strand is over the bigger slope strand at each crossing). For a Legendrian knot, the Thurston-Bennequin number (tb for short) is defined to be the writhe minus half of the number of cusps. In these special diagrams we understand that the framing on each 2–handle is one less than tb as in item (2) in Theorem 2.1. By abuse of language, a handle decomposition as in Theorem 2.1 is called a Stein structure.

Similarly for a given contact manifold (Y, ξ), one can attach 1–handles and $tb - 1$ framed 2–handles to $Y \times \{1\} \subset Y \times [0,1]$, in order to form a *Stein cobordism* built on (Y, ξ).

The embedding theorem relies on the fact that Stein manifolds are equivalent to *positive allowable Lefschetz fibrations* ($PALF$ for short). Recall that a $PALF$ on a 4–manifold W is a Lefschtetz fibration over a disk whose fibers have non-empty boundary and vanishing cycles are all non-separating curves. The restriction of a $PALF$ on the boundary $Y = \partial W$ is an open book decomposition whose monodromy can be written as a product of right handed Dehn twists. By [7], every Stein 4–manifold admits a $PALF$ (and every $PALF$ has a Stein structure). The construction is algorithmic where input is a handlebody decomposition of a Stein 4–manifold W as described in Theorem 2.1 and the output is a $PALF$ of W which is unique up to positive stabilization. Moreover, it is proved in [25] that the open book induced by this $PALF$ is compatible (in the sense of [14]) with the contact structure ξ induced by the Stein structure.

Given a Stein manifold W, fix a compatible $PALF$ as in the previous paragraph. Then one can extend this $PALF$ to a Lefschetz fibration over a closed manifold. Here is a sketch of what to do: First, recall the chain relation in the mapping class group of a surface. Let Σ_g be a surface of genus g. Let $\beta_1, \beta_2, \cdots, \beta_{2g}$ be a set of non-separating simple closed curves such that the following hold:

- $|\beta_i \cap \beta_j| = 1$ if $|i - j| = 1$,
- $|\beta_i \cap \beta_j| = 0$ if $|i - j| > 1$.

Proposition 2.2. *Let t_α denote the mapping class of the right handed Dehn twist about a curve α. Then following relation holds in the mapping class group of Σ_g:*

$$(t_{\beta_1} t_{\beta_2} \cdots t_{\beta_{2g}})^{4g+2} = 1. \tag{2.1}$$

Now on the Stein manifold W, we first choose a $PALF$ which induces an open book on $Y = \partial W$ with connected binding. This open book is compatible with the induced contact structure ξ. Then we attach a 2–handle along the binding with 0–framing relative to the page framing. Denote the corresponding cobordism by $V_0 : Y \to Y_0$. By [13], V_0 can be equipped with a symplectic structure extending the one defined on a collar neighborhood of Y. On the other hand, Y_0 is a surface bundle over the circle whose monodromy can be written as a product of right handed Dehn twists along non-separating curves. Let F denote a generic fiber in Y_0. Next, we use the chain relation of Proposition 2.2 to trivialize the monodromy by attaching -1 framed 2–handles: Write the monodromy of Y_0 as a product of right handed Dehn twists $t_{\gamma_1} \cdots t_{\gamma_n}$, where each γ_i is a non-separating curve on F. There is a diffeomorphism of F identifying γ_i with β_1 for each $i = 1, \cdots, n$. Using this diffeomorphism and the chain relation, we can write $t_{\gamma_i}^{-1}$ as a product of right handed Dehn twists. Attaching 2–handles as necessary, we can trivialize the monodromy. Finally we attach a copy $F \times D^2$ to get a cobordism $V_1 : Y_0 \to \emptyset$. Note that V_1 itself admits a Lefschetz fibration (with closed fibers) over disk. The closed 4–manifold $X := W \cup V_0 \cup V_1$ naturaly admits a Lefshcetz fibration over S^2. It is also possible to show that the Lefschetz fibration has a section, and hence symplectic, and that the construction can be made to guarantee that $b_2^+(X) \geq 2$. In other words, $V = V_0 \cup V_1$ is a *concave symplectic filling* for (Y, ξ). We summarize this construction in the following statement.

Theorem 2.3 ([7]). *Every Stein fillable contact manifold (Y, ξ) admits a concave symplectic filling $V = V_0 \cup V_1$, where V_0 is the cobordism $Y \to Y_0$ corresponding to a 2–handle attachment along the binding of an open book compatible with ξ, and V_1 admits a Lefschetz fibration over disk with closed fibers, which extends the fibration on Y_0. Moreover, V can be chosen in such a way that $b_2^+(V) \geq 2$.*

3. Heegaard Floer homology

Heegaard Floer homology ([21], [22]) is a type of Lagrangian Floer homology for the symmetric product of a Heegaard surface of a 3–manifold. There are several versions denoted by $\widehat{HF}(Y,\mathfrak{t})$, $HF^+(Y,\mathfrak{t})$, $HF^-(Y,\mathfrak{t})$, $HF^\infty(Y,\mathfrak{t})$. All of these groups are invariants of a 3-manifold Y with a Spin^c structure $\mathfrak{t}$. When $c_1(\mathfrak{t})$ is torsion these groups are $\mathbb{Q}$-graded. Each one admits an endomorphism U of degree -2 which makes all of them $\mathbb{Z}[U]$ modules (the action of U on $\widehat{HF}$ is trivial). They also satisfy the property that any Spin^c cobordism $(M,\mathfrak{s}) : (Y_1,\mathfrak{t}_1) \to (Y_2,\mathfrak{t}_2)$, from $(Y_1,\mathfrak{t}_1)$ to $(Y_2,\mathfrak{t}_2)$, induces a homomorphism

$$F^\circ_{(M,\mathfrak{s})} : HF^\circ(Y_1,\mathfrak{t}_1) \to HF^\circ(Y_2,\mathfrak{t}_2) \tag{3.1}$$

where HF° represents any of $\widehat{HF}, HF^+, HF^-$, or HF^∞. When both $c_1(\mathfrak{t}_1)$ and $c_1(\mathfrak{t}_2)$ are torsion, these homomorphisms shift degree by

$$d(M,\mathfrak{s}) = \frac{c_1(\mathfrak{s})^2 - 3\sigma(M) - 2\chi(M)}{4}. \tag{3.2}$$

These homomorphisms also satisfy the following *composition law*: given two Spin^c cobordisms $(M_1,\mathfrak{s}_1) : (Y_1,\mathfrak{t}_1) \to (Y_2,\mathfrak{t}_2)$ and $(M_2,\mathfrak{s}_2) : (Y_2,\mathfrak{t}_2) \to (Y_3,\mathfrak{t}_3)$, and $F^\circ_{M_1,\mathfrak{s}_1}$ and $F^\circ_{M_2,\mathfrak{s}_2}$ are the induced homomorphisms, their composition is given by

$$F^\circ_{M_2,\mathfrak{s}_2} \circ F^\circ_{M_1,\mathfrak{s}_1} = \sum_{\mathfrak{s} \in \mathrm{Spin}^c(M_1 \cup W_2) : \mathfrak{s}|_{M_i} = \mathfrak{s}_i} F^\circ_{M_1 \cup M_2, \mathfrak{s}}. \tag{3.3}$$

The following long exact sequence exists for every Spin^c 3–manifold $(Y,\mathfrak{t})$ and it is natural under cobordism–induced homomorphisms.

$$\cdots \xrightarrow{\ \delta\ } HF^-(Y,\mathfrak{t}) \xrightarrow{\ \iota\ } HF^\infty(Y,\mathfrak{t}) \xrightarrow{\ \pi\ } HF^+(Y,\mathfrak{t}) \xrightarrow{\ \delta\ } \cdots \tag{3.4}$$

The connecting homomorphism δ is an isomorphim between $\mathrm{coker}(\pi)$ and $\ker \iota$. We denote these groups respectively by $HF^+_{\mathrm{red}}(Y,\mathfrak{t})$ and $HF^-_{\mathrm{red}}(Y,\mathfrak{t})$, and call both of them by the same name: the *reduced* Heegaard Floer homology.

Heegaard Floer homology also provides a 4-manifold invariant (c.f. [23]). To review its definition first recall the mixed homomorphisms. Let $(M,\mathfrak{s}) : (Y_1,\mathfrak{t}_1) \to (Y_2,\mathfrak{t}_2)$ be a Spin^c cobordism with $b_2^+(M) \geq 2$. Every such cobordism admits an admissible cut, i.e., M can be decomposed as a union of two codimension zero sub-manifolds M_1 and M_2 with $b_2^+(M_i) \geq 1$, $i = 1, 2$, and $\delta H^1(N) \subseteq H^2(M)$ is trivial where N is the common boundary of these two sub-manifolds. Let $\mathfrak{s}_i = \mathfrak{s}|_{M_i}$ and $\mathfrak{t} = \mathfrak{s}|_N$. Then we have the following commutative diagram.

$$
\begin{array}{ccccc}
HF^+(Y_1,\mathfrak{t}_1) & \xrightarrow{\ \delta\ } & HF^-(Y_1,\mathfrak{t}_1) & \xrightarrow{\ \iota\ } & HF^\infty(Y_1,\mathfrak{t}_1) \\[2pt]
\downarrow{\scriptstyle F^+_{M_1,\mathfrak{s}_1}} & & \downarrow{\scriptstyle F^-_{M_1,\mathfrak{s}_1}} & & \downarrow{\scriptstyle F^\infty_{M_1,\mathfrak{s}_1}=0} \\[2pt]
HF^+(N,\mathfrak{t}) & \xrightarrow{\ \delta\ } & HF^-(N,\mathfrak{t}) & \xrightarrow{\ \iota\ } & HF^\infty(N,\mathfrak{t}) \\[2pt]
\downarrow{\scriptstyle F^+_{M_2,\mathfrak{s}_2}} & & \downarrow{\scriptstyle F^-_{M_2,\mathfrak{s}_2}} & & \downarrow{\scriptstyle F^\infty_{M_2,\mathfrak{s}_2}=0} \\[2pt]
HF^+(Y_2,\mathfrak{t}_1) & \xrightarrow{\ \delta\ } & HF^-(Y_2,\mathfrak{t}_1) & \xrightarrow{\ \iota\ } & HF^\infty(Y_2,\mathfrak{t}_1)
\end{array}
$$

The mixed homomorphism

$$F^{\text{mix}}_{(M,\mathfrak{s})} : HF^-(Y_1, \mathfrak{t}_1) \to HF^+(Y_2, \mathfrak{t}_2)$$

is defined by the composition $F^+_{(M,\mathfrak{s}_2)} \circ \delta^{-1} \circ F^-_{(M,\mathfrak{s}_1)}$. In [23], it is proved that the mixed homomorphism is independent of the admissible cut.

To define the 4–manifold invariant, we need to recall the Heegaard-Floer homology groups of the 3–sphere S^3 with its unique Spin^c structure:

$$HF^+_n(S^3) = \begin{cases} \mathbb{Z} & \text{If n is even and } n \geq 0 \\ 0 & \text{If n is odd,} \end{cases}$$

$$HF^-_n(S^3) = \begin{cases} \mathbb{Z} & \text{If n is even and } n \leq -2 \\ 0 & \text{If n is odd,} \end{cases}$$

Now, let X be a closed 4–manifold with $b_2^+(X) \geq 2$ and $\mathfrak{s}$ be a Spin^c structure on X. For simplicity, assume that $b_1(X) = 0$. Puncture X at two points and regard it as a cobordism from the 3–sphere to itself. Let $\Theta^{\pm}_{(n)}$ denote the generator of $HF^{\pm}_n(S^3)$. The *Ozsváth-Szabó* 4–manifold invariant is a linear map $\Phi_{X,\mathfrak{s}} : \mathbb{Z}[U] \to \mathbb{Z}$ which is described as follows: $\Phi_{X,\mathfrak{s}}(U^n)$ is characterized uniquely by the formula

$$F^{\text{mix}}_{X,\mathfrak{s}}(U^n \Theta^-_{(-2)}) = (\Phi_{X,\mathfrak{s}}(U^n))\Theta^+_{(0)}.$$

The 4-manifold invariant is zero on elements of degree not equal to $d(X, \mathfrak{s})$. A Spin^c structure $\mathfrak{s}$ on X is called a *basic class* if $\Phi_{X,\mathfrak{s}} \neq 0$. Finding the set of all basic classes of a given 4-manifold is an important problem in low-dimensional topology. The *adjunction inequality* gives a very strong restriction on the set of basic classes.

Theorem 3.1 ([23]). *Let X be a closed 4–manifold. Let $\Sigma \subset X$ be a homologically non-trivial embedded surface with genus $g \geq 1$ and with non-negative self-intersection number. Then for each Spin^c structure $\mathfrak{s} \in \text{Spin}^c(X)$ for which $\Phi_{X,\mathfrak{s}} \neq 0$, we have that*

$$|\langle c_1(\mathfrak{s}), [\Sigma] \rangle| + [\Sigma] \cdot [\Sigma] \leq 2g - 2. \tag{3.5}$$

The following is another version of the adjunction inequality along with a non-vanishing result of the 4–manifold invariant for Lefschetz fibrations.

Theorem 3.2 ([24]). *Let $\pi : X \to S^2$ be a relatively minimal Lefschetz fibration over sphere with generic fiber F of genus $g > 1$, and $b_2^+ > 1$. Then for the canonical Spin^c structure $\mathfrak{s}$ the map $F^{\text{mix}}_{X,\mathfrak{s}}$ sends the generator of $HF^-_{-2}(S^3)$ to the generator of $HF^+_0(S^3)$ (and vanishes on the rest of $HF^-(S^3)$). In particular $\mathfrak{s}$ is a basic class. For any other Spin^c structure $\mathfrak{t} \neq \mathfrak{s}$ with $\langle c_1(\mathfrak{t}), [F] \rangle \leq 2 - 2g = \langle c_1(\mathfrak{s}), [F] \rangle$, the map $F^{\text{mix}}_{X,\mathfrak{t}}$ vanishes.*

Given a contact structure ξ on Y, let $\mathfrak{t}_\xi$ be the induced Spin^c structure. A Heegaard Floer (co-)homology class $c^+ \in HF^+(-Y, \mathfrak{t}_\xi)$, which is an invariant of the isotopy class of ξ, is constructed in [20] as follows: Take an adapted open book decomposition for ξ which has connected binding. One can always find such an open book by doing positive stabilizations to any adapted open book as necessary. Let Y_0 denote the result of the

0–surgery on the binding, and $V_0 : Y \to Y_0$ be the associated cobordism. Naturally, Y_0 admits a fibration over circle. Let $\mathfrak{t}_0$ the Spin^c structure corresponding to the tangent plane distribution of fibers.

Proposition 3.3 ([24]). $HF^+(-Y_0, \mathfrak{t}_0) = \mathbb{Z}.$

Let c be a generator of $HF^+(-Y_0, \mathfrak{t}_0)$. It can be shown that there is a unique extention $\mathfrak{s}$ of the Spin^c structure $\mathfrak{t}_0$ over the cobordism W. The contact invariant is defined to be the image of c under the homomorphism which is induced by the Spin^c cobordism $(V_0, \mathfrak{s})$.

Definition 3.4. $c^+(\xi) := F^+_{V_0, \mathfrak{s}}(c) \in HF^+(-Y, \mathfrak{t}_\xi)/(\pm 1).$

Note that we turned V_0 upside down in this construction. This invariant is independent of the choice of the adapted open book decomposition used in its definition.

By using this definition along with the adjunction inequality and the symplectic compactification theorem, it can be proven that the contact invariant of a Stein fillable contact structure is in the image of the mixed homomorphism of some concave filling (c.f. [25]). From Theorem 2.3, we know that every Stein fillable contact manifold (Y, ξ) admits a concave symplectic filling $V = V_0 \cup V_1$ where V_0 is a cap off cobordism and V_1 is a Lefschetz fibration over a disk. Let $\mathfrak{s}$ be the canonical Spin^c structure.

Lemma 3.5 ([25]). $F^{\mathrm{mix}}_{V, \mathfrak{s}}(\Theta^-_{(-2)}) = \pm c^+(\xi)$ *if* $c_1(\xi)$ *is torsion.*

This lemma will play a key role in our argument. Its proof relies on the special topology of the concave filling constructed in Theorem 2.3. One would hope to prove it for arbitrary concave fillings (with $b_2^+ \geq 2$), but the authors do not know how to do it in full generality.

We are going to need a variant of Lemma 3.5 where one is allowed to add any Stein cobordism to the concave filling.

Lemma 3.6. *Let* (Y, ξ) *be a Stein fillable contact manifold with torsion* $c_1(\xi)$. *Let* M *be any Stein cobordism built on* (Y, ξ) *which does not contain any 1–handles. Then* M *can be extended to a concave filling* V *of* (Y, ξ) *such that* $F^{\mathrm{mix}}_{V, \mathfrak{s}}(\Theta^-_{(-2)}) = \pm c^+(\xi).$

Proof. Let (Y_1, ξ_1) be the convex end of M. Take a Stein filling of (Y, ξ) and glue it to M in order to obtain a Stein filling of (Y_1, ξ_1). Pick a $PALF$ of this Stein manifold and apply the algorithm in the proof of the Theorem 2.3 to find a concave filling $V = V_0 \cup V_1$ of (Y_1, ξ_1). If we glue this to M, we get a concave filling V' of (Y, ξ). By changing the order of some 2–handle attachments we can write this as $V' = V_0 \cup V_1'$, where $V_1' = M \cup V_1$ which is a Lefschetz fibration on disk. Now apply Lemma 3.5. $\qquad\square$

It is possible to define a relative version of the Ozsváth–Szabó 4–manifold invariant in the presence of a contact structure on the boundary. For, let W be a 4–manifold with connected boundary and ξ be a contact structure on ∂W. Given a Spin^c structure $\mathfrak{s}$ on W, the relative invariant $\Phi_{W, \mathfrak{s}}(\xi) \in \mathbb{Z}/\pm 1$ is uniquely characterized by the formula

$$F^+_{W, \mathfrak{s}}(c^+(\xi)) = \Phi_{W, \mathfrak{s}}(\xi)\Theta^+_{(0)}.$$

Again we puncture W and turn it upside down to regard it as a cobordism from $-\partial W$ to S^3. A twisted version of this invariant is conjecturally equivalent to the relative Seiberg–Witten invariant defined in [17].

4. Main theorem

With all the necessary tools in hand, we are ready to prove our main result. Henceforth suppose that W is a cork corresponding to an admissible link L. Let ξ be the induced contact structure for some choice of a Stein structure on W. Let τ be the involution on ∂W obtained by exchanging the components of L. Let ξ' be the pull back contact structure $\tau^*\xi$.

Theorem 4.1. *The contact invariants* $c^+(\xi)$ *and* $c^+(\xi')$ *in* $HF^+(-\partial W)$ *are distinct. Moreover, both elements descend non-trivially to* $HF^+_{\mathrm{red}}(-\partial W)$.

Proof. The trick is to inflate the cork using a Stein handle so that the cork twist changes the framing on the handle. This trick was first used in [10] to generate exotic Stein manifold pairs. We start by attaching a 2-handle to ∂W along a trefoil with framing 1 as in the left hand side of Figure 2. Thanks to the non-trivial linking with the 1-handle, this handle attachment induces a Stein cobordism M built on $(\partial W, \xi)$. On the right hand side of the same figure, however, the handle attachment can not be realized as a Stein cobordism, because the maximum Thurston-Bennequin number of trefoil is 1.

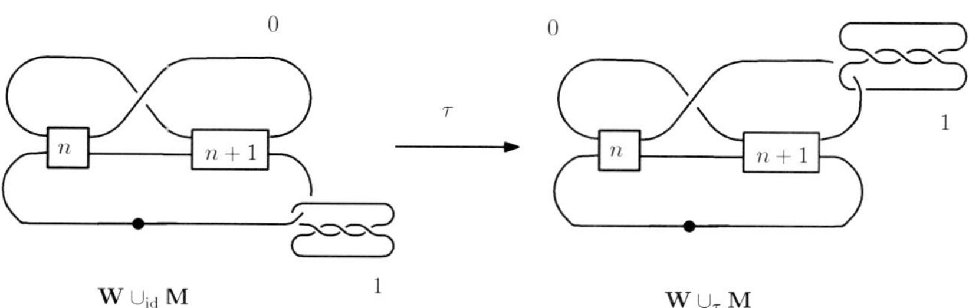

FIGURE 2

Next we apply Lemma 3.6 in order to extend M to a concave filling V of $(\partial W, \xi)$ whose mixed homomorphism hits the contact invariant $c^+(\xi)$.

The symplectic manifold $X := M \cup V$ admits a relatively minimal Lefschetz fibration. Let $\mathfrak{s}$ denote its canonical Spin^c structure. By Theorem 3.2, Lemma 3.6 and the compostion law, we have

$$\begin{aligned}
\Theta^+_{(0)} &= F^{\mathrm{mix}}_{W \cup V, \mathfrak{s}}(\Theta^-_{(-2)}) \\
&= F^+_{W,\mathfrak{s}} \circ F^{\mathrm{mix}}_{V,\mathfrak{s}}(\Theta^-_{(-2)}) \\
&= \pm F^+_{W,\mathfrak{s}}(c^+(\xi)).
\end{aligned}$$

In particular $F^+_{W,\mathfrak{s}}(c^+(\xi)) \neq 0$. On the other hand if we remove the cork W from X and reglue it using the cork twist τ, we obtain the manifold $X' := W \cup_\tau V$ which is homeomorphic to X. The right hand side of Figure 2 is $W \cup_\tau M$ and it embeds into X'. In that figure the trefoil represents an embedded torus of self intersection 1, which violates the adjunction inequality (Theorem 3.1). Therefore X' has no basic classes. This implies

$$\begin{aligned}
0 &= F^{\mathrm{mix}}_{W \cup_\tau V, \mathfrak{s}}(\Theta^-_{(-2)}) \\
&= F^+_{W,\mathfrak{s}} \circ \tau^* \circ F^{\mathrm{mix}}_{V,\mathfrak{s}}(\Theta^-_{(-2)}) \\
&= \pm F^+_{W,\mathfrak{s}}(\tau^*(c^+(\xi))).
\end{aligned}$$

This proves that $c^+(\xi)$ and $c^+(\xi') = \tau^*(c^+(\xi))$ are distinct. To prove the last statement, note that up to sign the U–equivariant involution $\tau^* : HF^+(-\partial W) \to HF^+(-\partial W)$ fixes the image of $HF^\infty(-\partial W)$ under the homomorphism π in Equation 3.4. Since $c^+(\xi)$ and $c^+(\xi')$ are not fixed by τ they descend non-trivially to $\mathrm{coker}(\pi)$. $\qquad\square$

The following corollary was previously proved in [4] by using different techniques.

Corollary 4.2. *The contact structures ξ and ξ' are homotopic, contactomorphic but not isotopic.*

Proof. We use the homotopy classification of 2–plane fields on a 3–manifold (see [15]). Since ∂W is an integral homology sphere, ξ and ξ' have the same two dimensional invariant. These two have also the same 3–dimensional invariant because they can be connected by a Stein cobordism which is topologically trivial: Simply take the symplectizations of $(\partial W, \xi)$ and $(\partial W, \xi')$, and glue two ends using τ. This proves that ξ and ξ' are homotopic. By definition τ defines a contactomorphism between these two contact structures. Theorem 4.1 shows that they are not isotopic. $\qquad\square$

Next corollary was first proved in [3] for the Mazur manifold, and in [11] in the general form.

Corollary 4.3. *The cork twist $\tau : \partial W \to \partial W$ does not extend inside of W as a diffeomorphism.*

Proof. Let $\mathfrak{s}$ be the unique Spin^c structure on W. Let ξ be the contact structure induced by a Stein structure. The proof of Theorem 4.1 shows that $F_{W,\mathfrak{s}}(c^+(\xi)) = \Theta^+_{(0)}$ and $F_{W,\mathfrak{s}}(\tau^* c^+(\xi)) = 0$. This shows that the relative invariants $\Phi_{W,\mathfrak{s}}(\xi)$ and $\Phi_{W,\mathfrak{s}}(\tau^*\xi)$ are different. $\qquad\square$

The following corollary was first proved in [5] by using adjunction inequality. Let us first introduce a terminology. Two Spin^c manifolds $(X, \mathfrak{s})$ and $(X', \mathfrak{s}')$ are said to be fake copies of each other if they are homeomorphic but not diffeomorphic.

Corollary 4.4. *The cork W can be symplectically embedded in some closed symplectic 4–manifold X so that removing W and regluing it via cork twist produces a fake copy X, with its canonical Spin^c structure.*

Proof. Pick a concave symplectic filling V of $(\partial W, \xi)$ as in Theorem 2.3. The manifold $X = W \cup V$ is simply connected and symplectic. Let $\mathfrak{s}$ be the canonical Spin^c structure of X. Now, remove the cork and reglue it using cork twist. The manifold $X' = W \cup_\tau V$ is simply connected and has the same intersection form as X. By Freedman's theorem there is a homeomorphism $f : X \to X'$. Let $\mathfrak{s}' = f_*(\mathfrak{s})$. We will prove that $\mathfrak{s}'$ is not a basic class. By Lemma 3.5

$$
\begin{aligned}
F^{\mathrm{mix}}_{X',\mathfrak{s}'}(\Theta^-_{(-2)}) &= F^+_{W,\mathfrak{s}} \circ \tau^* \circ F^{\mathrm{mix}}_{V,\mathfrak{s}}(\Theta^-_{(-2)}) \\
&= \pm F^+_{W,\mathfrak{s}}(\tau^*(c^+(\xi))) \\
&= 0.
\end{aligned}
$$

$\square$

Remark 4.5. *Note that if the inflated cork $W \cup_{id} M$ (of Theorem 1.1) embeds symplectically into a symplectic manifold X, then the cork twist produces a fake copy of X since in the cork twisted manifold the adjunction inequality fails. This is what is used in [5] and [11].*

Remark 4.6. *Let us formulate Corollary 4.4 in other terms: There is a concave filling V of $(\partial W, \xi)$ such that attaching V to W in two different ways produce two closed Spin^c 4–manifolds $(X, \mathfrak{s})$ and $(X', \mathfrak{s}')$ that are fake copies of each other. A large family of concave fillings satisfy this condition and presumably this also holds for all concave fillings (with $b_2^+(V) \geq 2$). Proving the latter, however, requires a generalization of Lemma 3.5 for arbitrary concave fillings. Once this is done, one would be able to show the following:*

Conjecture: If a cork embeds symplectically into any symplectic manifold X, such that $b_2^+(X) \geq 2$, then the cork twist produces a fake copy of X.

References

[1] S. Akbulut, On 2-dimensional homology classes of 4-manifolds, *Math. Proc. Camb. Phil. Soc.*, **82** (1977), 99–106.

[2] S. Akbulut, An involution permuting Floer Homology, *Turkish J. Math.*, **18** (1994), 16-22.

[3] S. Akbulut, A fake contractible 4–manifold, J. Diff. Geo., **33** (1991), 335–356

[4] S. Akbulut and R. Matveyev, A note on contact structures, *Pac. Jour of Math.*, **182** (1998), 201–204.

[5] S. Akbulut and R. Matveyev, Exotic structures and adjunction inequality, *Turkish Journal of Math.*, **21** (1997), 47–53.

[6] S. Akbulut and S. Durusoy, An involution acting non-trivially on Heegaard-Floer homology, *Fields Institute Communications*, **47** (2005), 1–9.

[7] S. Akbulut and B. Ozbagci, Lefschetz Fibrations on Compact Stein Surfaces, *Geometry and Topology*, **5** (2001), 319–334.

[8] S. Akbulut and B. Ozbagci, On the Topology of Compact Stein Surfaces, *Int. Math. Res. Notes*, **15** (2002), 769–782.

[9] S. Akbulut and K. Yasui, Corks, Plugs and exotic structures, *Jour. of GGT*, **2** (2008), 40–82.

[10] S. Akbulut and K. Yasui, Small exotic Stein manifolds, *Comm. Math. Helvetici*, **85** (2010), 705–721.

[11] S. Akbulut and K. Yasui, Stein 4–manifolds and corks, arXiv preprint, arXiv:1010.4122.

[12] Y. Eliashberg, Topological characterization of Stein manifolds in dimension > 2, *Int. Jour. of Math.*, **1** (1990), 29–46.

[13] Y. Eliashberg, A few remarks about symplectic fillings, *Geom. Topol.*, **8** (2004), 277–293.

[14] E. Giroux, Gomtrie de contact: de la dimension trois vers les dimensions suprieures, in Proceedings of the International Congress of Mathematicians, Vol. II (Beijing, 2002), 405–414, Higher Ed. Press, Beijing, 2002.

[15] R. Gompf, Handlebody contsruction of Stein surfaces, *Ann. of Math.*, **148** (1998), 619–693.

[16] R. Matveyev, A decomposition of smooth simply-connected h-cobordant 4-manifolds, *J. Diff. Geom.*, **44** (1996), 571–582.

[17] P. B. Kronheimer and T. S. Mrowka, Monopoles and contact structures, *Invent. Math.*, **130** (1997), 209-255.

[18] P. Lisca and G. Matić, Tight contact structures and Seiberg-Witten invariants, *Invent. Math.*, **129** (1997), 509-525.

[19] B. Ozbagci and A. Stipsicz, *Surgery on contact 3-manifolds and Stein surfaces.* Bolyai Society Mathematical Studies, 13. Springer-Verlag, Berlin; János Bolyai Mathematical Society, Budapest, 2004.

[20] P. Ozsváth and Z. Szabó, Heegaard Floer homology and contact structures, *Duke Math. J.*, **129** (2005), 39–61.

[21] P. Ozsváth and Z. Szabó, Holomorphic disks and topological invariants for closed three-manifolds, *Ann. of Math. (2)*, **159** (2004), 1027–1158.

[22] P. Ozsváth and Z. Szabó, Holomorphic disks and three-manifold invariants: properties and applications, *Ann. of Math. (2)*, **159** (2004), 1159–1245.

[23] P. Ozsváth and Z. Szabó, Holomorphic triangles and invariants for smooth four-manifolds, *Adv. in Math.*, **202** (2006), 326-400.

[24] P. Ozsváth and Z. Szabó, Holomorphic triangle invariants and the topology of symplectic four-manifolds, *Duke Math. J.*, **121** (2004), 1–34.

[25] O. Plamenevskaya, Contact structures with distinct Heegaard Floer invariants, *Math. Res. Let.*, **11** (2004), 547-561.

[26] N. Saveliev, A note on Akbulut Corks, *Math. Res. Lett.*, **10** (2003), 777-785.

DEPARTMENT OF MATHEMATICS MICHIGAN STATE UNIVERSITY, EAST LANSING 48824 MI, USA
E-mail address:　akbulut@math.msu.edu

DEPARTMENT OF MATHEMATICS THE UNIVERSITY OF TEXAS AT AUSTIN, 2515 SPEEDWAY, RLM 8.100 AUSTIN, TX 78712
E-mail address:　karakurt@math.utexas.edu

52

Proceedings of 18th Gökova
Geometry-Topology Conference
pp. 53 – 84

Lefschetz fibrations on cotangent bundles of two-manifolds

Joe Johns

ABSTRACT. In this paper our aim is to explain an explicit and relatively simple construction of some symplectic Lefschetz fibrations on the disk cotangent bundle of an arbitrary compact two dimensional manifold. Morally, the construction is inspired by the idea of complexifying a given Morse function on the 2-manifold. The paper is meant to give a treatment of a simple case in [J09], with extra focus on examples and concrete constructions.

1. Introduction

Let N denote a compact two dimensional manifold without boundary. Let $D(T^*N)$ denote the closed unit disk bundle of the cotangent bundle of N with respect to some metric on the bundle T^*N. We equip $D(T^*N)$ with the canonical 1-form θ which makes it into an exact symplectic manifold. The goal of this paper is to construct a Lefschetz fibration

$$\pi : D(T^*N) \longrightarrow D^2$$

with an explicit description of the regular fiber M and the vanishing cycles $V_1, \ldots, V_k$ in M.

The basic motivation is that we obtain an explicit presentation for the $4-$manifold $D(T^*N)$ in terms of the regular fiber M (which is a 2-manifold with boundary) and the vanishing cycles $V_1, \ldots, V_k$ (which are some circles in M). For the reader unfamiliar with these things, we explain in §2 the definition of a Lefschetz fibration, the regular fiber, and the vanishing cycles; and in §3 we explain two examples in some detail.

One way we might try to proceed is to start with a Morse function $f : N \longrightarrow \mathbb{R}$ and then "complexify" it in some sense to get a Lefschetz fibration $f_{\mathbb{C}} : D(T^*N) \longrightarrow \mathbb{C}$. For example, if $N = \mathbb{R}^2$, and $p : \mathbb{R}^2 \longrightarrow \mathbb{R}$ is a real polynomial which is a Morse function, then the obvious extension $p_{\mathbb{C}} : \mathbb{C}^2 \longrightarrow \mathbb{C}$ is a Lefschetz fibration on $\mathbb{C}^2 = T^*(\mathbb{R}^2)$ which we call the complexification of p. In this case, there is indeed a beautiful explicit description of the regular fiber and vanishing cycles of $p_{\mathbb{C}}$, given by A'Campo in [AC99].

Key words and phrases. Exact symplectic Lefschetz fibrations, cotangent bundles, 2-manifolds.
This work was supported by a scholarship at the Max Planck Institute, Leipzig, Germany.

This suggests a two step approach:

(i) Choose a Morse function $f : N \longrightarrow \mathbb{R}$, and complexify f in some sense to get a Lefschetz fibration $f_{\mathbb{C}} : D(T^*N) \longrightarrow \mathbb{C}$.

(ii) Describe the fiber M and vanishing cycles $V_1, \ldots, V_k$ of $f_{\mathbb{C}}$ in terms of the Morse theory of f.

Unfortunately, both steps (i) and (ii) seem to be difficult to carry out if N is an arbitrary 2-manifold. In step (i), it is difficult to define the "complexification" $f_{\mathbb{C}}$ so as to obtain Lefschetz fibration on $D(T^*N)$. In step (ii), even if we succeed in step (i) in some cases, it is not obvious how to describe the fiber and vanishing cycles explicitly.

In this paper, we will therefore take slightly different approach. Namely, we carry out the following four steps.

(1) We choose a Morse function f on N and a metric g such that (f, g) is Morse-Smale (which means the stable and unstable manifolds of $-\nabla_g f$ intersect transversely). Then we take the handle decomposition of N induced by (f, g).

(2) Given the handle decomposition of N, we give an explicit construction of an exact symplectic 2-manifold manifold M with contact boundary, and we specify some exact Lagrangian spheres (circles) $V_1, \ldots, V_k \subset M$. Here, there is one V_j for each critical point of f.

(3) We make use of a well-known construction which allows us to produce an exact symplectic Lefschetz fibration with any prescribed regular fiber and vanishing cycles (see Theorem 2.2). In this case we obtain a Lefschetz fibration

$$\pi : E \longrightarrow D^2$$

with regular fiber equal to M and with vanishing cycles $V_1, \ldots, V_k$. Here, E is some exact symplectic $4-$manifold with codimension 2 corners, which is determined by the construction.

(4) We prove that E is conformally exact symplectomorphic to the disk cotangent bundle $D(T^*N)$ (after we smooth the corners of E). See Theorem 1.1 below.

Thus, the key idea is to make an educated guess in step (2) for what the regular fiber M and vanishing cycles $V_1, \ldots, V_k$ ought to be. This guess is then verified to be correct in step (4). The source of inspiration for the construction in step (2) is A'Campo's paper [AC99] which we mentioned earlier, about complexifications of real polynomials.

The more precise result corresponding to step (4) is the following Theorem; the proof is sketched in §6.

Theorem 1.1. *Let N be a closed 2-manifold, equipped with a Morse function $f : N \longrightarrow \mathbb{R}$ and let g be a metric such that (f, g) is Morse-Smale. Let $\pi : E \longrightarrow D^2$ be a Lefschetz fibration constructed according to steps (1) to (3) above. Then (E, π) has the following properties:*

- *There is an exact Lagrangian embedding $N \subset E$.*
- *$Crit(\pi) \subset N$, $\pi(N) \subset \mathbb{R}$, and $\pi|N = f : N \longrightarrow \mathbb{R}$ (up to reparameterizing N and $\mathbb{R}$ by diffeomorphisms).*
- *E is conformally exact symplectomorphic to the disk cotangent bundle $D(T^*N)$ (after we smooth the corners of E).*

Remark 1.2. Steps (1) to (4), and Theorem 1.1, work perfectly well in the case where N has boundary, but to keep things simple we stick to the case where N is without boundary in this paper. But see §5.4 for some simple examples of how the construction works in the case where N has boundary.

Remark 1.3. We make some technical remarks. First, smoothing the corners of E is done in a standard way - see [S08] lemma 7.6. Second, we recall that a conformal exact symplectomorphism between exact symplectic manifolds is by definition a map

$$\phi : (E_1, \omega_1, \theta_1) \longrightarrow (E_2, \omega_2, \theta_2)$$

such that $\phi^*\theta_2 = \lambda\theta_1 + df$ for some smooth function $f : E_1 \longrightarrow \mathbb{R}$, and some $\lambda \in \mathbb{R}$ with $\lambda > 0$. (Here, $\omega_i = d\theta_i$ is an exact symplectic structure on E_i.) In particular $\phi^*\omega_2 = \lambda\omega_1$, so ϕ is a conformal symplectomorphism. The basic example of a conformal exact symplectomorphism is given by integrating the Liouville vector field of T^*N to give a map from a neighborhood of N in T^*N onto a smaller neighborhood of N in T^*N, with strictly less symplectic volume.

In some cases it *is* possible to carry out steps (i) and (ii), and in such cases it is natural to compare the result with the Lefschetz fibrations constructed as in steps (1) to (4). In §3 we carry out steps (i) and (ii) for $D(T^*\mathbb{R}P^2)$ and $D(T^*T^2)$ (in these examples, we can make use of the embeddings $D(T^*\mathbb{R}P^2) \subset \mathbb{C}P^2$ and $D(T^*T^2) \subset (\mathbb{C}^*)^2$). In §5 we explain how to construct the regular fiber vanishing cycles as in step (2) for $D(T^*T^2)$ and $D(T^*\mathbb{R}P^2)$. Interestingly, we find the *same* regular fiber and vanishing cycles that we got via steps (i) and (ii) for $D(T^*T^2)$ and $D(T^*\mathbb{R}P^2)$ in §3. In general, it remains an interesting problem to carry out steps (i) and (ii) for more general manifolds and compare the result to our construction in steps (1) to (4).

We conclude the introduction with a quick summary of the contents of the paper. Note that §2 and §3 are of a more pedagogical nature, and they play only a peripheral role in the rest of the paper.

In §2 we give the definition of a Lefschetz fibration, and review some of the basic theory. In §3 we explain two examples of Lefschetz fibrations, on $D(T^*\mathbb{R}P^2)$ and $D(T^*T^2)$, which come from algebraic geometry. In §4 we explain step (2) above. That is, we describe how to construct M and $V_1, \ldots, V_k \subset M$ for an arbitrary closed 2-manifold N. We also specify the basis of vanishing paths we want to use. In §5 we illustrate how the construction of M and $V_1, \ldots, V_k \subset M$ works in some examples, in particular, for the cases $N = \mathbb{R}P^2$ and $N = T^2$. In §6 we sketch the proof of Theorem 1.1.

2. A quick lesson on Lefschetz fibrations and Picard-Lefschetz theory

In this section we discuss some of the basic theory of exact symplectic Lefschetz fibrations. As general references we recommend [AGZV88, GS99, L81] for a classical point of view on Lefschetz fibrations and [S08, S03A, AS04] for a symplectic point of view. We remark that §2.8 describes the main result on constructing Lefschetz fibrations that will be used later in the paper.

2.1. The definition of an exact symplectic Lefschetz fibration

Let M be a manifold with boundary. An exact symplectic structure on M is a $1-$form θ such that $\omega = d\theta$ is a symplectic form; and in addition the Liouville vector field X_θ defined by $\omega(X_\theta, \cdot) = \theta$ must be transverse to the boundary of M and point outwards; this makes the boundary into a contact manifold, using θ as a contact form.

The simplest example of an exact symplectic Lefschetz fibration is a trivial fiber bundle

$$\pi : E = M \times D^2 \longrightarrow D^2$$

where the fiber is given by an exact symplectic manifold M, and the base $D^2 \subset \mathbb{C}$ has the standard exact symplectic structure. Roughly speaking, a nontrivial Lefschetz fibration is similar except it is allowed to have finitely many *singular fibers* of a certain type. That is, there are finitely many points $p_1, \ldots, p_k \in Int(E)$ which are isolated critical points of π of Morse type. We give the precise definition in moment. For now we just observe that the total space of an exact symplectic Lefschetz fibration is naturally a manifold with codimension 2 corners, because the fiber and base both have boundary.

Let E be a manifold with codimension 2 corners. An exact symplectic structure on E is a $1-$form θ such that $\omega = d\theta$ is a symplectic form, and such that the Liouville vector field X_θ is transverse to each boundary stratum of codimension 1, and points outwards; this makes each boundary stratum into a contact manifold, using θ as a contact form.

Let E be an exact symplectic manifold with codimension 2 corners. An exact symplectic Lefschetz fibration on E is a map

$$\pi : E \longrightarrow D^2$$

such that the following conditions are satisfied (following [S08]):

- There are finitely many points $p_1, \ldots, p_k \in Int(E)$ such that, for each p_j, there are complex coordinates $(z_1, \ldots, z_n)$ near p_j such that p_j corresponds to 0; the standard complex structure $J_0 = i$ on $\mathbb{C}^n$ is compatible with ω; and

$$\pi = z_1^2 + \ldots + z_n^2 + \pi(p_j)$$

in these coordinates. The points p_j are called the *singular points or critical points* of π; this type of singularity is called a Morse type singularity.

- For each $z \in D^2 \setminus \{c_1, \ldots, c_k\}$, where $c_j = \pi(p_j)$, we require that $M_z = \pi^{-1}(z)$ is an exact symplectic submanifold of E, where $\theta_z = \theta|M_z$ makes it into an exact symplectic submanifold with boundary. For each c_j, $\pi^{-1}(c_j) \setminus \{p_j\}$ is a (noncompact) exact symplectic manifold with boundary.
- Set $\partial_v E = \pi^{-1}(\partial D^2)$ and let $\partial_h E$ denote the boundary of all the fibers (including singular ones). Then we require that $\partial E = \partial_h E \cup \partial_v E$, where ∂E has two boundary strata $\partial_h E$ (the horizontal part) and $\partial_v E$ (the vertical part) which meet at a codimension 2 corner. Furthermore, $\pi|\partial_h E \longrightarrow \partial D^2$ and $\pi|\partial_h E \longrightarrow D^2$ are required to be surjective fiber bundles.
- For each $x \in E$ consider the splitting

$$T_x E = T_x^h E \oplus T_x^v E,$$

where $T_x^v E = Ker(D\pi_x)$ and $T_x^h E$ is the symplectic complement in $T_x E$. This splitting always exists; our requirement is

$$T_x^h E = T_x(\partial_h E) \text{ for all } x \in \partial_h E.$$

2.2. Parallel transport

The purpose of the last condition in the definition of a Lefschetz fibration (we drop the adjective "exact symplectic" from now on) is to ensure that parallel transport is well-defined, as follows. Set $M_z = \pi^{-1}(z)$ for any $z \in D^2$. Given any two regular values $x, y \in D^2 \setminus \{c_1, \ldots, c_k\}$, and any path $\gamma : [0,1] \longrightarrow D^2 \setminus \{c_1, \ldots, c_k\}$ from x to y which avoids the critical values, we define a *parallel transport map*

$$\tau_\gamma : M_x \longrightarrow M_y$$

by using the connection given by $T^h E$. The last condition ensures that the vector field given by the horizontal lifts of $\gamma'(t)$ for each t can be integrated for all $t \in [0,1]$ and that τ_γ maps the boundaries of the fibers into themselves.

2.3. The regular fiber

Let us now give a more intuitive description of a Lefschetz fibration. As we saw above, if x, y are two regular values τ_γ gives an exact symplectic isomorphism from M_x to M_y. Thus, a Lefschetz fibration is, roughly speaking, a fiber-bundle with one common fiber M, except that finitely many of the fibers have isolated singular points (modeled locally by $z_1^2 + \ldots + z_n^2 = 0$ in $\mathbb{C}^n$). Thus we speak of "the regular fiber" M, which is defined to be $M = \pi^{-1}(b)$ for some fixed regular value b.

Remark 2.1. The basic reason Lefschetz fibrations on E are useful for studying E is that they give rise to a dimensional reduction: One can shift focus from E to the regular fiber M which is two dimensions less (and then, potentially, one can repeat this process for M and so on ...). So, for example, if E is a 4-manifold, we can instead focus on a 2-manifold M. In principle, this philosophy may be applicable to many different questions about the symplectic topology of E, such as the symplectomorphism type (classification and construction of exotic structures, see [M09, MS09, AS10]), the symplectomorphism group of E and M (see [AMP05, S06]), and Lagrangian submanifolds of E and M (see [S03B, S08, FSS08]).

2.4. Vanishing paths

For the rest of this section we fix some notation, in order to discuss some general properties of Lefschetz fibrations.

- Let $\pi : E \longrightarrow D^2$ be a Lefschetz fibration.
- Let $p_1, \ldots, p_k \in E$ denote the critical points of π.
- Let $c_1, \ldots, c_k \in Int(D^2)$ denote the critical values of π.
- Fix a base point $b \in D^2 \setminus \{c_1, \ldots, c_k\}$.
- Set $M = \pi^{-1}(b)$, set $M_x = \pi^{-1}(x)$, for any $x \in D^2 \setminus \{c_1, \ldots, c_k\}$, and set $M_j = \pi^{-1}(c_j)$ for all j.
- For each j we pick a path $\gamma_j : [0,1] \longrightarrow D^2$ from b to c_j which avoids all other $c_i \neq c_j$. And assume that $\gamma_i((0,1]) \cap \gamma_j((0,1]) = \emptyset$ for all $i \neq j$.

If $(\gamma_1, \ldots, \gamma_k)$ is a collection of paths satisfying the last condition, we call it a *basis of vanishing paths* for (E, π).

2.5. Vanishing cycles and Lefschetz thimbles

Associated to each vanishing path γ_j there is an exact Lagrangian sphere $V_j = V_{\gamma_j}$ in the regular fiber M. This is called the *vanishing cycle*, or vanishing sphere, associated to γ_j. Roughly speaking, if we follow the parallel transport map along γ_j from M to the singular fiber M_j then V_j collapses down to a point in M_j, which is the singular point p_j. This gives a nice description of each singular fiber M_j.

More precisely, we have the following lemma (for a proof see [S03A] §1). Since $T_x^h E = 0$ for $x = p_j$, we first have to say what we mean by the transport map $\tau_{\gamma_j} : M \longrightarrow M_j$. We define

$$\tau_{\gamma_j}(x) = \lim_{t \longrightarrow 1} \tau_{\gamma_j|_{[0,t]}}(x).$$

Lemma 1. Let $\pi : E \longrightarrow D^2$ be a Lefschetz fibration with critical points $p_1, \ldots, p_k$. Let $(\gamma_1, \ldots, \gamma_k)$ be a basis of vanishing paths. Set $M = \pi^{-1}(b)$ and $M_j = \pi^{-1}(c_j)$ for each j. Then for each j there is an exact Lagrangian sphere $V_j \subset M$ such that the parallel transport map

$$\tau_{\gamma_j} : M \longrightarrow M_j$$

is such that

$$\tau_{\gamma_j}|_{M \setminus V_j} : M \setminus V_j \longrightarrow M_j \setminus \{p_j\},$$

is an exact symplectic isomorphism, and satisfies $\tau_{\gamma_j}(V_j) = \{p_j\}$. Moreover, V_j comes equipped with a diffeomorphism

$$\phi_j : S^{n-1} \longrightarrow V_j$$

which is determined as a well-defined element $[\phi_j] \in \pi_0(\mathrm{Diff}(S^{n-1})/O(n))$.

To each vanishing path γ_j there is also an associated Lagrangian disk

$$\Delta_j = \Delta_{\gamma_j} \subset Int(E),$$

called the Lefschetz thimble of γ_j. Roughly speaking, Δ_j is the trace of the vanishing cycle V_j as it is transported over the path γ_j and eventually collapses to the critical point p_j. More precisely, for $t \in [0,1)$, let $V_j(t)$ denote the vanishing cycle in $\pi^{-1}(\gamma_j(t))$ corresponding to the restricted path $\gamma_j|_{[t,0]}$ (so $V_j(1) = V_j$, in particular). Then

$$\Delta_j = \left(\bigcup_{t \in [0,1)} V_j(t) \right) \cup \{p_j\}.$$

In particular Δ_j has the following properties: $\pi(\Delta_j) = \gamma_j([0,1])$, $\partial\Delta_j = V_j$, and p_j is in $Int(\Delta_j)$. We give one more perspective: If $\gamma_j([0,1]) \subset \mathbb{R}$ and $\gamma_j : [0,1] \longrightarrow \mathbb{R}$ is an embedding, then we can consider the Morse function $f = Re(\pi) : Int(E) \longrightarrow \mathbb{R}$ and Δ_j is just the unstable manifold of $\pm\nabla_g f$ at $p_j \in Crit(f)$ with respect to any metric g on $Int(E)$. (More precisely, Δ_j is the part of the unstable manifold of p_j lying in $f^{-1}([a,b])$, where $[a,b] = \gamma_j([0,1])$.)

2.6. An example: the standard local model

The simplest example of a Lefschetz fibration with at least one critical point comes from the map $q : \mathbb{C}^n \longrightarrow \mathbb{C}$, where

$$q(z_1, \ldots, z_n) = z_1^2 + \ldots + z_n^2.$$

This example is of course very important because it serves as the local model near any critical point of an arbitrary Lefschetz fibration. We now summarize some key features of q. For a detailed discussion we refer to [S03A], §1.

Every regular fiber $q^{-1}(z)$ is exact symplectomorphic to T^*S^{n-1}. To see that, we note that for any $s > 0$,

$$q^{-1}(s) = \{x + iy \in \mathbb{C}^n : |x|^2 - |y|^2 = s, x \cdot y = 0\};$$

and we realize T^*S^n as

$$T^*S^{n-1} = \{(u,v) \in \mathbb{R}^n \times \mathbb{R}^n : |u| = 1, u \cdot v = 0\}$$

equipped with the restriction of the standard exact symplectic structure on $\mathbb{R}^n \times \mathbb{R}^n \cong \mathbb{C}^n$. Now we define a map

$$\sigma_s : q^{-1}(s) \longrightarrow T^*S^{n-1}$$

by the formula $\sigma_s(x + iy) = (u,v)$, where $u = \frac{x}{|x|}$, $v = -|x|y$; this is an exact symplectomorphism.

To make q into an exact symplectic Lefschetz fibration, we cut down the fiber from

T^*S^{n-1} to the disk bundle $D(T^*S^{n-1})$, and we cut down the base from $\mathbb{C}$ to D^2, as follows. For $z \in \mathbb{C}^n$, set

$$k(z) = \frac{1}{4}(|z|^4 - |q(z)|^2).$$

Then $k \geq 0$ and the sub-level sets of k precisely cut down the fibers of q to disk bundles with respect to the standard metric:

$$\sigma_s(q^{-1}(s) \cap \{k \leq r\}) = D_r(T^*S^{n-1}).$$

Let $r, s > 0$ and set

$$E_{r,s} = \{z \in \mathbb{C}^n : |q(z)| \leq s, |k(z)| \leq r\}, \ \pi_{r,s} = q|E_{r,s}.$$

Then it is easy to see that the symplectic complement to $Ker(Dq_z)$ is precisely $\bar{z}\mathbb{C} \subset \mathbb{C}^n = T_z(\mathbb{C}^n)$, and that Dk_z is zero on $\bar{z}\mathbb{C}$. Thus

$$\partial_h E_{r,s} = \{k = r\} \cap E_{r,s} \text{ and } \partial_v E_{r,s} = \{|q| = s\} \cap E_{r,s}$$

and $T_x^h(E) = T_x(\partial_h E)$ for $x \in \partial_h E$, because dk is zero on $T_x^h(E)$.

Take $s = 1$ and set $E_0 = E_{r,1}$, $\pi_0 = \pi_{r,1}$. Then (E_0, π_0) is a Lefschetz fibration with regular fiber isomorphic to $D_r(T^*S^{n-1}) = \{(u, v) \in T^*S^{n-1} : |v| \leq r\}$. The vanishing cycle can be described as follows. For *any* path γ in D^2 from $b = 1$ to $c = 0$, the vanishing cycle V_γ is given by the zero section $S^{n-1} \subset D_r(T^*S^{n-1})$ (under the isomorphism $\sigma_1 : \pi_0^{-1}(1) \longrightarrow D_r(T^*S^{n-1})$).

Because an arbitrary Lefschetz fibration $\pi : E \longrightarrow D^2$ has the same local form given by $\pi_0 : E_0 \longrightarrow D^2$ near every critical point, we obtain the same picture for π near each critical point: A neighborhood of the critical point in E corresponds to a neighborhood of the origin in $\mathbb{C}^n$ (where the neighborhood of $0 \in \mathbb{C}^n$ can be taken to be $E_{r,s} \subset \mathbb{C}^n$) and a neighborhood of each vanishing cycle V_j in M corresponds to a neighborhood of S^{n-1} in T^*S^{n-1}.

2.7. Monodromy

The main classical result about Lefschetz fibrations is the Picard-Lefschetz theorem. We will not need this result, so we will not give the precise statement, but the rough idea is the following. Take a loop λ_j in D from b to b which winds counter-clockwise around c_j, and suppose λ_j does not wind around any $c_i \neq c_j$. Then the Picard-Lefschetz theorem asserts that the monodromy map $\tau_{\lambda_j} : M \longrightarrow M$ is isotopic to a Dehn twist around the vanishing sphere $V_j \subset M$. If $dim M = 2$, then V_j is a circle, and a Dehn twist is the familiar map from geometric topology. If $dim M > 2$ there is a generalization of the notion of Dehn twist in any symplectic manifold. See [S08] §16c for more details. Since $\pi_1(D^2 \setminus \{c_1, \ldots, c_k\})$ is generated by $\lambda_1, \ldots, \lambda_k$, the Picard-Lefschetz theorem can also be used to describe the monodromy map $\tau_\gamma : M \longrightarrow M$ up to isotopy, for any loop γ from b

to b which avoids $c_1, \ldots, c_k$. The corresponding map

$$\Theta : \pi_1(D^2 \setminus \{c_1, \ldots, c_k\}) \longrightarrow \pi_0(Symp(M))$$

is called the monodromy homomorphism.

2.8. Constructing Lefschetz fibrations

Given a Lefschetz fibration $\pi : E \longrightarrow M$, we call the data

$$(M, V_1, \ldots, V_k, \gamma_1, \ldots, \gamma_k),$$

Picard-Lefschetz data for (E, π), where $(\gamma_1, \ldots, \gamma_k)$ is a basis of vanishing paths, and $(V_1, \ldots, V_k)$ is the family of parameterized vanishing spheres determined by $(\gamma_1, \ldots, \gamma_k)$. Here, each V_j is *parameterized*, which means that each V_j comes with a diffeomorphism $\phi_j : S^{n-1} \longrightarrow V_j$, more precisely, an element $[\phi_j] \in \pi_0(\mathrm{Diff}(S^{n-1})/O(n))$, as in lemma 1.

In this paper, the main result we will need is the following theorem that we quote from [S08, lemma 16.9]. It says that any desired Picard-Lefschetz data can be realized by some exact symplectic Lefschetz fibration; and in fact there is an explicit construction of the desired Lefschetz fibration. The more precise statement goes as follows:

Theorem 2.2. *Let M be any exact symplectic manifold, and let $V_1, \ldots, V_k$ be any choice of parameterized exact Lagrangian spheres in M. Let $b \in D^2$, and let $c_1, \ldots, c_k \in Int(D^2)$ be any points with $b \neq c_j$ for all j. Let $(\gamma_1, \ldots, \gamma_k)$ be any choice of paths $\gamma_j : [0,1] \longrightarrow D^2$ satisfying the conditions of a basis of vanishing paths, that is: $\gamma_j(0) = b, \gamma_j(1) = c_j$, $\gamma_j(t) \neq c_i$ for all $t \in [0,1]$, and for all $i \neq j$, and $\gamma_i((0,1]) \cap \gamma_j((0,1]) = \emptyset$ for all $i \neq j$. Then, there exists a Lefschetz fibration*

$$\pi : E \longrightarrow D^2$$

equipped with a canonical isomorphism $\pi^{-1}(b) \cong M$ such that π has critical values $c_1, \ldots, c_k$ and the vanishing cycles corresponding to $(\gamma_1, \ldots, \gamma_k)$ are precisely $(V_1, \ldots, V_k)$, under the identification $\pi^{-1}(b) \cong M$.

The proof of this theorem follows from an explicit construction. The basic idea is start with the trivial fibration $M \times D^2 \longrightarrow D^2$ and cut and paste in the local model $\pi_0 : E_0 \longrightarrow D^2$ to produce a Lefschetz fibration E_j with exactly one vanishing cycle, $j = 1, \ldots, k$. Then we fiber-connect sum each $E_1, \ldots, E_k$ onto one more copy of the trivial fibration to get a Lefschetz fibration $\pi : E \longrightarrow D^2$.

We remark that there is also a definition of Lefschetz fibrations $\pi : X \longrightarrow \mathbb{C}P^1$, where the total space and fiber are closed symplectic manifolds, as in [AS04] for example. But there is no analogue of Theorem 2.2 in that setting.

3. Two examples of complexifications in algebraic geometry

In this section we discuss two examples of complexifications in algebraic geometry. The first is a Lefschetz fibration

$$\pi : D(T^*\mathbb{R}P^2) \longrightarrow D^2$$

which arises from the embedding $\mathbb{R}P^2 \subset \mathbb{C}P^2$. The second example is a Lefschetz fibration

$$\pi : D(T^*T^2) \longrightarrow D^2$$

which arises from the embedding $T^2 = S^1 \times S^1 \subset \mathbb{C}^* \times \mathbb{C}^*$.

3.1. A complexification of a Morse function $f : \mathbb{R}P^2 \longrightarrow \mathbb{R}$ using a classical Lefschetz pencil on $\mathbb{C}P^2$

The first example will be based on a Lefschetz fibration on $D(T^*\mathbb{R}P^2)$ using the algebraic geometry of a simple Lefschetz pencil on $\mathbb{C}P^2$. (This is a modification of the example in [AS04, p. 39].) Let $s_0 = x_1^2 + x_2^2$ and $s_1 = x_0^2 - x_2^2$ be two real homogeneous polynomials of degree 2. For each $\alpha \in \mathbb{C}$ we consider the subsets of $\mathbb{C}P^2$

$$C_\alpha = \{s_1 + \alpha s_0 = 0\}, \quad \text{and} \quad C_\infty = \{s_0 = 0\}.$$

For $\alpha \in \mathbb{C}$, we have $s_1 + \alpha s_0 = 0$ iff $\frac{s_1}{s_0} = -\alpha$ iff $[s_0, s_1] = [1, -\alpha]$ in $\mathbb{C}P^1$. Thus we can think of C_α as a family of subsets parameterized by $\alpha \in \mathbb{C}P^1 = \mathbb{C} \cup \{\infty\}$, where $\alpha = \infty$ just means $C_\infty = \{s_0 = 0\}$.

The collection C_α, $\alpha \in \mathbb{C}P^1$ is called a *Lefschetz pencil* of curves in $\mathbb{C}P^2$. We call C_α the "fiber" of the pencil lying above $\alpha \in \mathbb{C}P^1$. (The word "pencil" is a bad translation of the French word "pinceau", meaning "brush".)

For each α, if C_α is smooth then it follows from standard facts in algebraic geometry relating degree to Euler characteristic that C_α is diffeomorphic to S^2. If all the curves C_α were smooth, and if they were mutually disjoint, then they would fiber $\mathbb{C}P^2$ into a fiber bundle $\mathbb{C}P^2 \longrightarrow \mathbb{C}P^1$ with fiber S^2. However, this picture is not correct for two reasons: first, not all C_α are smooth, and so not all are diffeomorphic to S^2; and second, the curves C_α are not mutually disjoint. Let us discuss the second issue first.

The *base locus* of the pencil is by definition

$$B = \{s_0 = s_1 = 0\}$$

which in this case consists of four points

$$B = \{[1, 1, i], [1, -1, i], [1, 1, -i], [1, -1, -i]\}$$

The meaning of B is that any two of our curves C_α and $C_{\alpha'}$ meet in B.

Next, there are three singular fibers, each of which is the union of two lines in $\mathbb{C}P^1$:

$$C_\infty = \{s_0 = 0\} = \{x_1 = -ix_2\} \cup \{x_1 = ix_2\}$$

$$C_0 = \{s_1 = 0\} = \{x_0 = x_1\} \cup \{x_0 = -x_1\}$$

$$C_1 = \{s_1 + s_0 = 0\} = \{x_0 = -ix_1\} \cup \{x_0 = ix_1\}.$$

It is straight-forward to check that for $\alpha \notin \{0, 1, \infty\}$, that C_α is regular. (For this let $F_\alpha = s_1 + \alpha s_0$ and check that the complex rank of DF_α is 1 on C_α in each of the three standard charts $\psi_0(y_1, y_2) = [1, y_1, y_2]$, etc.)

Now we delete one regular curve from $\mathbb{C}P^2$, say $C_i = \{s_1 + is_0 = 0\}$. Then what remains,

$$X = \mathbb{C}P^2 \setminus C_i,$$

is the union of a family of curves $C_\alpha \setminus B$, $\alpha \in \mathbb{C}P^1 \setminus \{i\}$ which are mutually disjoint. There are three singular fibers and every regular fiber is diffeomorphic to a four times punctures sphere $S^2 \setminus \{\text{four points}\}$. We define

$$\pi : X \longrightarrow \mathbb{C}P^1 \setminus \{i\} \cong \mathbb{C}$$

to be the map which sends $p \in C_\alpha$ to α. Thus

$$\pi = -\frac{s_1}{s_0} : X \longrightarrow \mathbb{C}P^1 \setminus \{i\} \cong \mathbb{C}.$$

This defines a *Lefschetz fibration* on X in the sense of classical complex algebraic geometry. (This also yields an exact symplectic Lefschetz fibration with codimension 2 corners if we restrict π to a compact subset of X in such a way that we down the base to D^2 and we cut down the regular fiber to $S^2 \setminus \{$ four small open disks $\}$.)

Notice that our three singular fibers are related to the four base points in the following way. Take the line through two of the base points and take the other line through the other two base points. The union of these two lines yields one of the singular fibers. There are three ways to do this and in this way we obtain all three singular fibers.

Now, this actually tells us what the three vanishing cycles corresponding to the three singular fibers look like, as follows. Fix one regular fiber, which is $S^2 \setminus B$. Now take the singular fiber consisting of the line through $p_1, p_2 \in B$ and the line through $q_1, q_2 \in B \setminus \{p_1, p_2\}$. Then the corresponding vanishing cycle $V \subset S^2 \setminus B$ must be such that when we collapse V to a point the result consists of $S^2 \setminus \{p_1, p_2\}$ and $S^2 \setminus \{q_1, q_2\}$ meeting at one point. Thus, V must have divided B into two halves $\{p_1, p_2\}$ and $\{q_1, q_2\}$. There are therefore three vanishing cycles which divide B into pairs in all three possible ways. See figure 1.

Now consider $\mathbb{R}P^2 \subset \mathbb{C}P^2$. Notice that three singular fibers have critical points (where the two lines intersect) $[1, 0, 0]$, $[0, 1, 0]$, and $[0, 0, 1]$ which all lie in $\mathbb{R}P^2$. Also notice that $\pi(\mathbb{R}P^2) \subset \mathbb{R}$ and in fact $f = \pi|_{\mathbb{R}P^2}$ is the standard Morse function on $\mathbb{R}P^2$ with three critical points.

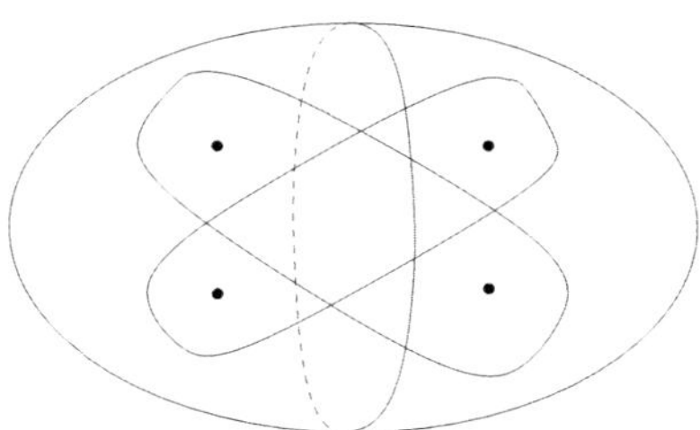

FIGURE 1. The regular fiber of the pencil $\pi = s_1/s_0$ (with the four base points deleted) and the three vanishing spheres.

Let $B_\epsilon(i)$ denote a small open disk around i such that $\epsilon < 1/2$ so that $\pi(\mathbb{R}P^2) \subset \mathbb{R}$ has a neighborhood disjoint from $B_\epsilon(i)$. Now set

$$D = \mathbb{C}P^1 \setminus B_\epsilon(i) \cong D^2$$

so that D is diffeomorphic to a compact disk and it contains $\pi(\mathbb{R}P^2)$ in its interior. Now instead of deleting just C_i from $\mathbb{C}P^2$ let us delete $U = \pi^{-1}(B_\epsilon(i))$, which we view as a small neighborhood of the fiber C_i which is disjoint from $\mathbb{R}P^2 \subset \mathbb{C}P^2$. Set

$$X_0 = \mathbb{C}P^2 \setminus U, \text{ and } \pi_0 = \pi|_{X_0} : X_0 \longrightarrow D.$$

Now for $\alpha \in D$, the fiber $\pi_0^{-1}(\alpha) = C_\alpha \setminus U$ consists of S^2 minus a small neighborhood of B, which means we get

$$\pi_0^{-1}(\alpha) = S^2 \setminus \{\text{four small disks}\}$$

with the same vanishing cycles.

In fact X is symplectomorphic to $D(T^*\mathbb{R}P^2)$ so that $\pi_0 : X_0 \longrightarrow D$ is a Lefschetz fibration on $D(T^*\mathbb{R}P^2)$ which we can view as a complexification (extension, in particular) of the standard Morse function $f : \mathbb{R}P^2 \longrightarrow \mathbb{R}$.

3.2. A complexification of a Morse function $f : T^2 \longrightarrow \mathbb{R}$ using a Laurent polynomial on $(\mathbb{C}^*)^2$

Realize the torus as

$$T^2 = S^1 \times S^1 \subset (\mathbb{C}^*)^2$$

Now define a Lefschetz fibration

$$W : (\mathbb{C}^*)^2 \longrightarrow \mathbb{C}$$

by the formula

$$W = x + \frac{1}{x} + y + \frac{1}{y}.$$

One can easily check this has four critical points $(x, y) = (\pm 1, \pm 1)$ and the complex Hessian $D^2 W$ is nondegenerate there; thus W is indeed a Lefschetz fibration.

Notice that
$$W|_{T^2} = 2\cos\theta_1 + 2\cos\theta_2$$
where we parameterize T^2 by
$$x = \cos\theta_1 + i\sin\theta_1,$$
$$y = \cos\theta_2 + i\sin\theta_2,$$
$$\theta_1, \theta_2 \in \mathbb{R}/2\pi\mathbb{Z}.$$
Thus $f = W|_{T^2}$ is the standard Morse function on T^2 (f is the sum of two copies of the height function on S^1). f has four critical points given by $\theta_1, \theta_2 \in \{0, \pi\}$, that is, $(x, y) = (\pm 1, \pm 1) \in S^1 \times S^1$. Thus all the critical points of W lie on $T^2 \subset (\mathbb{C}^*)^2$.

To describe the topology of the regular fiber of W, notice that
$$W(-1, -1) = -4, \quad W(1, 1) = 4, \quad W(-1, 1) = W(1, -1) = 0$$
so $b = 1$ is a regular value. Let us take the equation
$$x + \frac{1}{x} + y + \frac{1}{y} = 1$$
and multiply by xy to get
$$x^2 y + y + y^2 x + x - xy = 0.$$
Now let us homogenize this equation in the coordinates $[x, y, z] \in \mathbb{C}P^2$ to get an equation in $\mathbb{C}P^2$:
$$M = \{x^2 y + yz^2 + y^2 x + xz^2 - xyz = 0\} \subset \mathbb{C}P^2$$
Since this is a regular cubic in $\mathbb{C}P^2$ standard results in algebraic geometry relating the Euler characteristic to the degree tell us that
$$M \cong T^2.$$
Now to put M back into $(\mathbb{C}^*)^2$ we should delete the lines $x = 0$, $y = 0$, and $z = 0$. We note that
$$x = 0 \text{ implies } yz^2 = 0, \text{ so } y = 0 \text{ or } z = 0;$$
$$y = 0 \text{ implies } xz^2 = 0, \text{ so } x = 0 \text{ or } z = 0;$$
$$z = 0 \text{ implies } x^2 y = y^2 x, \text{ so } y = 0, \text{ or } x = 0, \text{ or } y = x.$$
Thus the intersection of M with the three lines consists of four points
$$(M \cap \{x = 0\}) \cup (M \cap \{y = 0\}) \cup (M \cap \{z = 0\})$$
$$= \{[0, 0, 1], [0, 1, 0], [1, 0, 0], [1, 1, 0]\}.$$
Thus the fiber of W is diffeomorphic to T^2 minus four points. We do not determine the vanishing cycles. But in §5.2 we shall construct a Lefschetz fibration
$$\pi : D(T^*T^2) \longrightarrow D^2$$
which has four critical points all lying on $T^2 \subset D(T^*T^2)$ and such that the restriction $\pi|_{T^2} = f : T^2 \longrightarrow \mathbb{R}$ is the standard Morse function above. The fiber of π is T^2 minus

four small disks (which is compact version of the fiber of W). Thus the natural conjecture is that the vanishing cycles of W coincide with those of π.

4. The construction of the fiber and vanishing cycles

In this section we explain how to carry out step (2) in our four step plan in the introduction.

4.1. The construction of the fiber M

Let N be a compact 2-manifold without boundary and let $f : N \longrightarrow \mathbb{R}$ be a Morse function. Take a metric g such that (f, g) is Morse-Smale. Assume that f has just one maximum and minimum and denote the critical points by x_0, x_1^j, x_2, where the subscript indicates the Morse index. (This assumption is for convenience; see section 4.3 for a discussion of how to remove this assumption if desired.) Take the handle-decomposition of N induced by (f, g). Let us denote the handles by h_0, h_1^j, h_2, where the subscript indicates the index.

The fiber M will be constructed as follows. First let $V_0 = S^1$. And set

$$M_0 = D(T^* V_0).$$

This can be thought of as the first approximation to our fiber; it contains only one vanishing cycle, namely V_0, which we think of as the vanishing cycle corresponding to x_0.

To finish constructing the fiber M we will attach $2k$ 1-handles to the boundary of

$$M_0 = D(T^* V_0) \cong S^1 \times [-1, 1].$$

Remark 4.1. As usual, the manifold resulting from each handle attachment must be smoothed afterwards to get a smooth manifold with boundary. We are working with exact symplectic 2-manifolds, so in our case we use the method of handle-attachments described by Weinstein [W91], so that the manifold resulting from attaching a handle has a canonical exact symplectic structure inherited form the one on $D(T^* V_0)$ plus the exact symplectic structure on the handle (there is a standard model of a handle in this context).

For each 1-handle h_1^j in the handle decomposition of N, we take two 1-handles $\tilde{h}_1^j$ and $\hat{h}_1^j$ which will be attached to $D(T^* V_0)$.

Each 1-handle h_1^j is attached to $h_0 = D^2$ at a pair of points

$$\{a_j, b_j\} \subset \partial h_0 = S^1,$$

called the attaching sphere (0-sphere) of h_1^j. Now we identify

$$V_0 = \partial h_0 = S^1.$$

In this way V_0 inherits the attaching spheres $\{a_j, b_j\} \subset V_0$, $j = 1, \ldots, k$. For each j we will attach attach two 1-handles, $\tilde{h}_1^j$ and $\hat{h}_1^j$, to the following four points in the boundary of $D(T^*V_0) \cong S^1 \times [-1, 1]$:

$$\{(-1, a_j), (-1, b_j), (1, a_j), (1, b_j)\} \subset \partial(S^1 \times [-1, 1]) \cong \partial(D(T^*V_0)).$$

The way the handles are attached will depend on the framing used to attach h_1^j to h_0, as follows. (One thing to keep in mind is the result M is supposed to be a symplectic 2-manifold, so it must be orientable.)

(1) First consider the case where h_1^j is attached to h_0 *with a twist* (for example, there is a twist in the 1-handle of the standard handle decomposition of $\mathbb{R}P^2$). In this case $\tilde{h}_1^j$ will be attached to $(a_j, 1)$ and $(b_j, 1)$ while $\hat{h}_1^j$ will be attached to $(a_j, -1)$ and $(b_j, -1)$. For both of the handles, the framing will be such that there is *no twist* in the handle. This means, more precisely, that there are small tubular neighborhoods

$$[a_j - \epsilon, a_j + \epsilon] \times \{1\} \text{ and } [b_j - \epsilon, b_j + \epsilon] \times \{1\}$$

such that the handle attachment is equivalent to identifying these neighborhoods while *preserving* the orientations of the intervals. (And the same statement holds with 1 replaced by -1.)

(2) Next consider the case where h_1^j is attached to h_0 *without a twist* (for example, both 1-handles in the standard handle decomposition of T^2 have no twists). In this case $\tilde{h}_1^j$ will be attached to $(a_j, 1)$ and $(b_j, -1)$ while $\hat{h}_1^j$ will be attached to $(a_j, -1)$ and $(b_j, 1)$. For both of the handles, the framing will be such that there is *no twist* in the handle. This means, more precisely, that there are small tubular neighborhoods

$$[a_j - \epsilon, a_j + \epsilon] \times \{1\} \quad \text{and} \quad [b_j - \epsilon, b_j + \epsilon] \times \{-1\}$$

such that the handle attachment is equivalent to identifying these neighborhoods while *preserving* the orientations of the intervals. (And the same statement holds with 1 and -1 exchanged.)

Once we have attached these $2k$ 1-handles to $D(T^*V_0)$, this completes the construction of the fiber, M.

Remark 4.2. Here is an alternative way of thinking of the construction of M which is conceptually appealing. Start with $D(T^*V_0)$ as before. Now for each $1-$handle h_1^j take a copy of the circle, denoted by

$$V_1^j = S^1.$$

This will be the vanishing cycle corresponding to the critical point x_1^j. Now for each j, do a *plumbing* of $D(T^*V_0)$ and $D(T^*V_1^j)$ such that in the plumbing

$$V_0 \cap V_1^j = \{a_j, b_j\}$$

and such that the result is the same as attaching the two 1-handles $\tilde{h}_1^j$ and $\hat{h}_1^j$ as before. See for example figure 3 for how this plumbing looks like. The conceptual advantage of plumbing is that the vanishing cycles V_1^j corresponding to x_1^j are apparent from the beginning.

4.2. Construction of the vanishing cycles

We have constructed our fiber M. It remains to describe the vanishing cycles in M. First, V_0, the vanishing cycle corresponding to x_0, is apparent. Next we define V_1^j, the vanishing cycle corresponding to x_1^j, to be union of the two segments:

$$(\{a_j\} \times [-1,1]) \cup (\{b_j\} \times [-1,1]) \subset V_0 \times [-1,1] \cong D(T^*V_0)$$

and the core of $\tilde{h}_1^j$ and the core of $\hat{h}_1^j$. See figures 3 and 5 for example. (Alternatively, see remark 4.2 at the end of the last section for another construction of M where V_1^j is quite manifest.)

The last vanishing cycle V_2 corresponding to the critical point x_2 is the most interesting one. It is obtained by doing the Lagrangian surgery of V_0 with all V_1^j simultaneously for $j = 1, \ldots, k$ (see [P91] for a general discussion of Lagrangian surgery). There is also a choice of "left" or "right" Lagrangian surgery; we have chosen a "left" surgery $V_0 \# V_1^j$, which means as we move along V_0 towards the surgery point the surgered curve moves to the left, see figure 4.

The reason we choose the left surgery is because of our choice of vanishing paths which we describe now. Our choice of left surgery corresponds to the fact that our vanishing paths (see figure 2) lie in the lower half plane $\{z \in \mathbb{C} : Re(z) \leq 0\}$. (If the vanishing paths were reflected to the upper half plane then the right surgery would have been appropriate in the definition of V_2). We choose the critical values in the unit disk $D^2 \subset \mathbb{C}$ to lie on the real line:

$$c_2 = 3/4, \ c_1 = 0, \ c_0 = -3/4.$$

And we choose the base point to be

$$b = -1/2.$$

Let γ_0, $\gamma_1 = \gamma_1^j$, $j = 1, \ldots, k$, and γ_2 denote the vanishing paths as in figure 2. Note that we chose all γ_1^j, $j = 1, \ldots, k$ to be all the same, equal to γ_1:

$$\gamma_1^1 = \ldots = \gamma_1^k = \gamma_1.$$

This is not a problem since the vanishing cycles V_1^j, $j = 1, \ldots, k$ are mutually disjoint.

4.3. General Morse functions

In this section we briefly discuss how to remove the assumption that there is only one critical point index 0, and one of index 2. (This discussion is not essential for the rest of the paper, so the reader is free to skip it, or come back to it later.) In the general case

FIGURE 2. Here we have the vanishing paths γ_0, $\gamma_1 = \gamma_1^j$, $j = 1, \ldots, k$, and γ_2 in $\mathbb{C}$. Here, γ_0 goes along the interval $[-3/4, -1/2]$ from $b = -1/2$ to $c_0 = -3/4$; γ_1 goes along the interval $[-1/2, 0]$ from $b = -1/2$ to $c_1 = 0$; and γ_2 goes from $b = -1/2$ to $c_2 = 3/4$ by going around $c_1 = 0$ via an arc in the lower half plane, and then along the interval $[1/2, 3/4]$.

we proceed as follows. Let $x_0^1, \ldots x_0^l$ denote the critical points of index 0; let $x_1^1, \ldots x_1^k$ denote the critical points of index 1; and let $x_2^1, \ldots x_2^m$ denote the critical points of index 2. Corresponding to this we have the handle decomposition of f with 0-handles h_0^i, 1-handles h_1^j, and 2-handles h_2^p, $i = 1, \ldots, l$, $j = 1, \ldots, k$, $p = 1, \ldots, m$.

To construct M, we start with l disjoint copies of $D(T^*S^1)$, denoted $D(T^*V_0^i)$, $i = 1, \ldots, l$. So we set $M_0 = \cup_i D(T^*V_0^i)$ as our first approximation to M. Now we identify each V_0^i with the boundary of h_0^i, that is $V_0^i = \partial h_0^i$. In this way $\cup_i V_0^i$ inherits several embedded copies of S^0, say $K^j = S^0$, $j = 1, \ldots k$, which are the attaching spheres of the 1-handles h_1^j, $j = 1, \ldots, k$. As before, each K_j gives rise to four points in the boundary of $\cup_i D(T^*V_0^i)$. That is, each pair of points $a_j, b_j \in V_0^i = S^1$ gives rise to $(a_j, \pm 1)$ and $(b_j, \pm 1)$ in $D(T^*V_0^i) = S^1 \times [-1, 1]$. To construct M we attach a pair of 1-handles $\widetilde{h}_1^j$, $\widehat{h}_1^j$ at each quadruple of points $(a_j, \pm 1)$, $(b_j, \pm 1)$ according to the same rules as before (that is, rule (1) or (2) in §4.1, depending whether the 1-handle h_1^j was attached with or without a twist in the handle decomposition of N).

The vanishing cycles V_0^i and V_1^j are evident as before. Now V_2^p, $p = 1, \ldots, m$ arise by doing the (left) Lagrangian surgery of $\cup_i V_0^i$ and $\cup_j V_1^j$ in M at each point where they meet. There will result from all these surgeries precisely m circles which are by definition V_2^p, $p = 1, \ldots, m$.

The reason we will obtain the correct number of circles (i.e., m) is that the Lagrangian surgery of $\cup_i V_0^i$ and $\cup_j V_1^j$ in M perfectly mirrors the surgery that happens in N as we pass between level sets,

from the level set $\partial[\cup_i h_0^i]$, to the level set $\partial[(\cup_i h_0^i) \cup (\cup_j h_1^j)]$.

As an example/exercise, we suggest considering $N = S^2$, where $f : S^2 \longrightarrow \mathbb{R}$ has two maxima, two minima, and two critical points of index 1. Before trying this exercise the reader may first wish to look at the next two sections, where we will carry out the examples $N = \mathbb{R}P^2$ and $N = T^2$ where the Morse function has only one maximum and one minimum in each case.

5. Some examples of the construction

In this section we explain how the construction described in §4 works in some examples, in particular, we consider the cases $N = T^2$ and $N = \mathbb{R}P^2$.

5.1. Example 1: $N = \mathbb{R}P^2$.

First, we consider the example $N = \mathbb{R}P^2$. In this case we have a handle decomposition with three handles h_0, h_1, h_2. To construct the fiber M of our Lefschetz fibration $\pi : D(T^*\mathbb{R}P^2) \longrightarrow D^2$, we start with $D(T^*V_0)$ and then attach two 1-handles $\tilde{h}_1$ and $\hat{h}_1$ to $D(T^*V_0)$. To begin, identify $\partial h_0 = S^1 = V_0$ and let $\{a, b\}$ denote the two points where h_1 is attached to h_0 (i.e., $\{a, b\}$ is the attaching 0-sphere). Thus we obtain two points $\{a, b\} \subset V_0$. Now identify

$$D(T^*V_0) = V_0 \times [-1, 1]$$

and consider the following four points in the boundary of $D(T^*V_0)$:

$$\{(a, -1)(a, 1), (b, 1), (b, -1)\} \subset \partial(V_0 \times [-1, 1]).$$

In the handle decomposition of $N = \mathbb{R}P^2$ the 1-handle h_1 is attached with a twist. Therefore, according to the instructions in §4.1 (1), we must attach $\tilde{h}_1$ to $(a, 1), (b, 1)$ and $\hat{h}_1$ to $(a, -1), (b, -1)$. The result is shown in figure 3 (notice that the 1-handles are attached without twists so that M is orientable).

There are three vanishing cycles. Two, V_0 and V_1 are quite obvious: In figure 3, V_0 is the horizontal segment, and V_1 is the circle which meets V_0 vertically; V_1 is the core of the evident annulus plumbed on to $D(T^*V_0)$.

Remark 5.1. Notice that V_0 and V_1 intersect in two points. These two points correspond exactly to the two flow lines from x_1 to x_0, which are represented by the two attaching points, a, b, where h_1 attached to h_0.

The last vanishing cycle V_2 is the Lagrangian surgery of V_0 and V_1. It is shown in figure 4: V_2 coincides with V_0 and V_1 away from a neighborhood of $V_0 \cap V_1$, and inside a neighborhood of $V_0 \cap V_1$ it coincides with the four curved arcs in figure 4.

Notice that the resulting fiber M is diffeomorphic to S^2 with four small disks removed. This agrees with the example we did from classical algebraic geometry in §3.1. Furthermore, V_0, V_1 and V_2 divide the four punctures in all three possible ways. Thus the vanishing cycles also agree with the example from classical algebraic geometry.

5.2. Example 2: $N = T^2$.

Next, we consider the example of the torus, $N = T^2$. In this case we have a handle decomposition with four handles h_0, h_1^1, h_1^2, h_2. To construct the fiber M of our Lefschetz fibration $\pi : D(T^*T^2) \longrightarrow D^2$, we start with $D(T^*V_0)$ and then attach four 1-handles to $D(T^*V_0)$ in pairs, denoted $\tilde{h}_1^1, \hat{h}_1^1$ and $\tilde{h}_1^2, \hat{h}_1^2$. To begin, consider the points $\{a_j, b_j\} \subset \partial h_0$

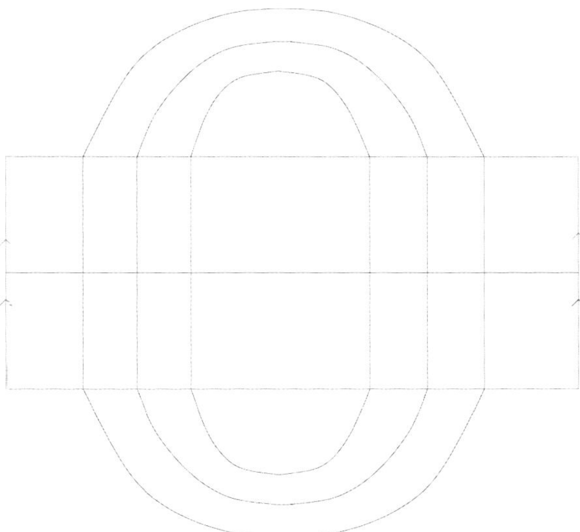

FIGURE 3. This is the fiber M in the case $N = \mathbb{R}P^2$. The two vertical edges marked by $>>$ are identified. The vanishing cycle V_0 is represented by the horizontal line in the middle. The vanishing cycle V_1 is the circle meeting V_0 vertically. The last vanishing cycle V_2 is not shown (for V_2 see figure 4).

where h_1^j is attached to h_0, $j = 1, 2$. Now identify $S^1 = V_0 = \partial h_0$. Thus we obtain four points in V_0:

$$a_1, b_1, a_2, b_2 \subset V_0.$$

The cyclic order in V_0 is: a_1, a_2, b_1, b_2, which we see from the handle decomposition of T^2. Now identify $D(T^*V_0) = V_0 \times [-1, 1]$ and for $j = 1, 2$, consider the following four points in the boundary of $D(T^*V_0)$:

$$\{(a_j, -1)(a_j, 1), (b_j, 1), (b_j, -1)\} \subset \partial(V_0 \times [-1, 1]).$$

In the handle decomposition of $N = T^2$ both 1-handles h_1^j, $j = 1, 2$ are attached with no twist. Therefore, according to the instructions in §4.1 (2), we must attach $\tilde{h}_1^j$ to $(a_j, 1), (b_j, -1)$ and $\hat{h}_1^j$ to $(a_j, 1), (b_j, -1)$. The result is shown in figure 5 (notice that the 1-handles are attached without twists so that M is orientable).

There are four vanishing cycles V_0, V_1^1, V_1^2, V_2. Three of them, V_0, V_1^1, and V_1^2 are quite obvious, see figure 5.

FIGURE 4. This is the fiber M in the case $N = \mathbb{R}P^2$. The two vertical edges marked by $>>$ are identified. The vanishing cycle V_2 is shown (it is curvy); it is the Lagrangian surgery of V_0 and V_1 which are shown in figure 3.

Remark 5.2. Notice that V_0 and V_1^j intersect in two points. These two points correspond exactly to the two flow lines from x_1^j to x_0, which are represented by the two attaching points, a_j, b_j, where h_1^j is attached to h_0.

The last vanishing cycle V_2 is the Lagrangian surgery of V_0 and $V_1^1 \cup V_1^2$. It is shown in figure 6: Outside of a neighborhood of the four intersection points,

$$(V_0 \cap V_1^2) \cup (V_0 \cap V_1^2) = \{a_1, b_1, a_2, b_2\},$$

V_2 coincides with V_0, V_1^1, and V_2; and inside a neighborhood of $\{a_1, b_1, a_2, b_2\}$, V_2 coincides with the eight curved arcs shown in figure 6. It is instructive to trace through the figure and verify that we indeed get a copy of S^1, which is V_2 by definition.

By inspecting the boundary carefully we see there are 4 boundary components. The Euler characteristic is -4 (to see this, we can retract M onto the 1-skeleton, which has one 0-cell and five 1-cells). Thus we conclude M is diffeomorphic to T^2 with 4 small disks removed. (Here we use the formula for the Euler characteristic of a compact 2-manifold Σ given by $\chi(\Sigma) = (2 - 2g) - b$, where $g = genus(\Sigma)$, and b is the number of boundary components of Σ.)

Thus M agrees with the fiber in the example of the Lefschetz fibration on $(\mathbb{C}^*)^2 \cong T^*T^2$ we

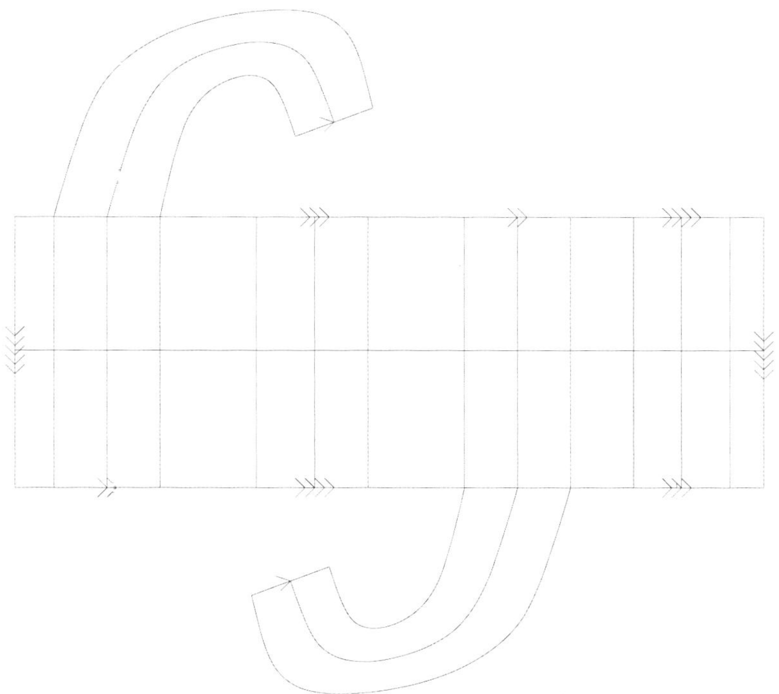

FIGURE 5. The fiber M in the case $N = T^2$. The two vertical edges indicated by $>>>>>$ are identified to form $S^1 \times [-1, 1] = D(T^* V_0)$. The other marked edges, in pairs, indicate where the four $1-$handles are attached. We have drawn in part of one of the 1-handles (up to identifying $>$ and $>$) to indicate how a handle would look. Three of the four vanishing cycles are visible: V_0 is the horizontal segment, and V_1^1 is represented by two of the vertical segments (joining the edges marked $>$ and $>>$) and V_1^2 is represented by the other two vertical segments (joining the edges marked $>>>$ and $>>>>$). Because of the identifications indicated, these vertical segments join together to form two copies of S^1. See figure 6 for the final vanishing cycle V_2.

considered in §3.2. In that example we had four critical points all lying on $(S^1)^2 \subset (\mathbb{C}^*)^2$ which corresponds to $T^2 \subset T^* T^2$. This agrees with the fact that we have four vanishing cycles in M. The natural conjecture is that the vanishing cycles of the example in §3.2 are the same as the ones in M.

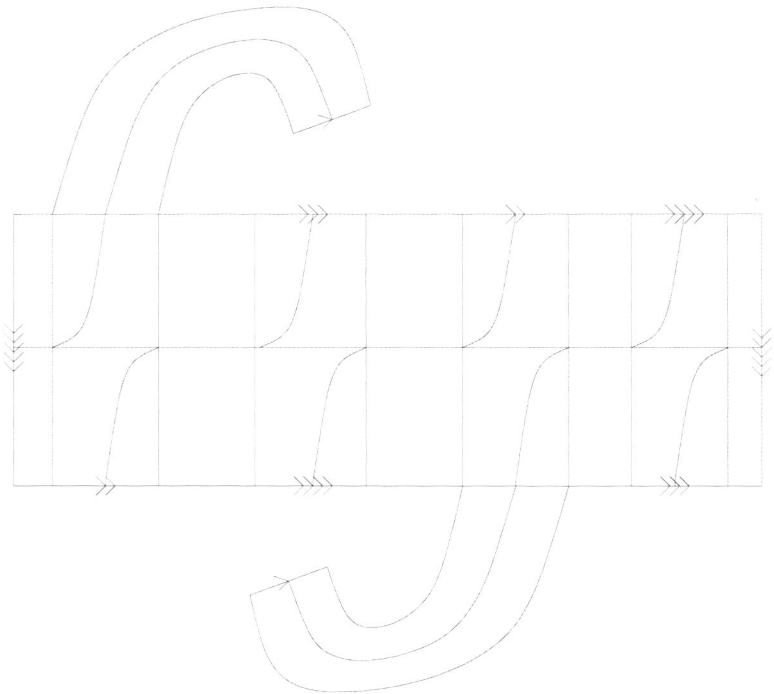

FIGURE 6. The fiber M in the case $N = T^2$. Here V_2 is shown; it is the Lagrangian surgery of V_0 and V_1^2, V_1^2. See figure 5 for a picture of V_0 and V_1^2, V_1^2.

5.3. The genus of the fiber in general

Let N be the closed oriented surface of genus g. Take the standard handle decomposition of N with one $0-$handle, one $2-$handle, and $2g$ $1-$handles. We describe without proof the fiber M in this case: If we construct M as in §4.1, and reason in a way similar to the example of the torus, then we find the fiber M is equal to a genus g surface with $2 + 2g$ small disks removed, and with $2g + 2$ vanishing cycles. Theorem 1.1 implies there is a Lefschetz fibration on $D(T^*N)$ with this regular fiber and vanishing cycles. This answer for M makes sense in view of the general fact that the total space of a Lefschetz fibration on a 4-manifold is homotopy equivalent to the fiber M with a 2-cell attached at each vanishing cycle. Indeed, in this case M is equal to N with $2g + 2$ small disks removed, so it is plausible that attaching $(2g + 2)$ 2-cells to the vanishing cycles should recover N as expected. (This is not a complete argument that E is homotopy equivalent to N because the the position of the vanishing cycles has not taken into account.)

5.4. Two quick examples where N has boundary

Let
$$N_1 = T^2 \setminus D,$$
where D is an embedded disk, and let
$$N_2 = \mathbb{R}P^2 \setminus D',$$
where D' is an embedded disk. Then N_1 and N_2 admit Morse functions and handle decompositions which are the same as the ones we had in §5.1, 5.2 except we remove a neighborhood of the maximum, and correspondingly we omit h_2 from the handle decompositions. In these cases the regular fibers are exactly the same as before, as in §5.1, 5.2, and we have the same vanishing cycles, except we omit V_2 in both cases.

6. Sketch of $E \cong D(T^*N)$

Let N be a 2-manifold and construct M and $V_0, V_1^j, V_2 \subset M$ as described in §4. Take the vanishing paths $\gamma_0, \gamma_1^j, \gamma_2$ as described in §4.2. Now we invoke Theorem 2.2 and obtain a Lefschetz fibration
$$\pi : E \longrightarrow D^2$$
which, by construction, has the Picard-Lefschetz data
$$(M, V_0, V_1^j, V_2, \gamma_0, \gamma_1^j, \gamma_2).$$

In this section we sketch the proof of the following theorem:

Theorem 6.1.   (1) *There is an exact Lagrangian embedding $N \subset E$.*
  (2) *$Crit(\pi) \subset N$, $\pi(N) = [a, b] \subset \mathbb{R}$, and $\pi|N = f : N \longrightarrow \mathbb{R}$ (up to reparameterizing N and $\mathbb{R}$ by diffeomorphisms).*
  (3) *E is conformally exact symplectomorphic to the disk cotangent bundle $D(T^*N)$ (after we smooth the corners of E).*

The main step is (1). Then (2) follows easily by construction of N. The last step (3) is accomplished by describing a retraction of E onto a small Weinstein neighborhood of N in E, symplectomorphic to $D(T^*N)$. The retraction is obtained by using the parallel transport map with varying time along some fixed paths; that is why it is a conformally exact symplectomorphism. See remark 1.3 for more about conformally exact symplectomorphisms. As a model example one should imagine some neighborhood of the zero section in T^*N retracting onto the disk bundle $D_\epsilon(T^*N)$ for some small $\epsilon > 0$ via the Liouville flow along the cotangent fibers, with varying time. This is obviously not symplectic because it distorts the symplectic volume, but it is a conformal exact symplectomorphism, which is the best one could hope for in this situation, and still useful.

We will discuss the proof in the case $N = \mathbb{R}P^2$, as in §5.1. The proof in the general case involves no new ideas. But see remark 6.2 for a brief discussion of the general case.

The regular fiber M and vanishing spheres V_0, V_1, V_2 are as in figure 8 below. The critical values in D^2 are chosen to be

$$\begin{aligned}
\pi(x_2) &= c_2 = -3/4, \\
\pi(x_1) &= c_1 = 0, \\
\pi(x_0) &= c_0 = 3/4.
\end{aligned}$$

And the base point is

$$b = -1/2.$$

Let γ_0, γ_1, γ_2 denote the vanishing paths in figure 2.

Now let $b' = 1/2$ and consider the obvious reflection of the vanishing paths through the origin, γ_0', γ_1', γ_2'. Then let

$$M' = \pi^{-1}(b').$$

We will suppress how M and M' are identified. (Roughly speaking they are related by a "half twist operation", and then transport around the half circle from $1/2$ to $-1/2$ gives an exact symplectomorphism from M' to M which can be understood explicitly. See Lemma 6.1 and Lemma 7.2 in [J09] for details.)

If we identify $M = M'$ then the vanishing spheres V_0', V_1', V_2' in M' appear as in figure 7. Under the identification $M' = M$, we have $V_1' = V_1$, $V_2' = V_0$ and V_0' looks similar to V_2, but the surgery is the *right* surgery of V_0' and V_1', which means as you move along V_0' towards the surgery region, the curve moves to the right. Compare with figure 8 which shows $V_0, V_1, V_2 \subset M$.

At this point we need to go into the construction of (E, π) in a bit more detail (see [J09] for full details). Let $(E_{r,s}, \pi_{r,s})$ denote the standard local model (see §2.6) with fiber $D_r(T^* S^n)$, $r > 0$; but use $q_1 = z_1^2 - z_2^2 + c_1$ rather than the usual $q = z_1^2 + z_2^2$. Let

$$\phi_1 : D_r(T^* S^1) \longrightarrow M$$

denote an exact Lagrangian embedding such that $\phi_1(S^1) = V_1$ (such an embedding exists by Weinstein's theorem).

Let E_1 denote the Lefschetz fibration over $D_s(c_1) = \{z \in \mathbb{C} : |z - c_1| \leq s\}$ obtained in the following way. Take the trivial fibration

$$F = M \times D_s(c_1),$$

and consider the subset obtained by deleting a neighborhood of V_1 in every fiber:

$$F_0 = (M \setminus \phi_1(D_{r/2}(T^* S^1))) \times D_s(c_1).$$

76

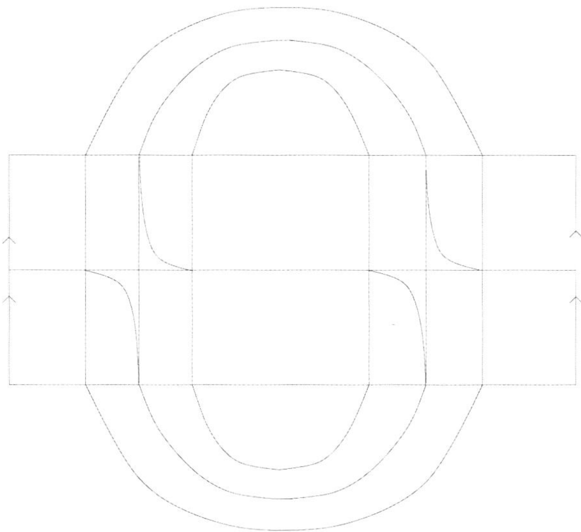

FIGURE 7. In the case $N = \mathbb{R}P^2$: The fiber M' at $b' = 1/2$

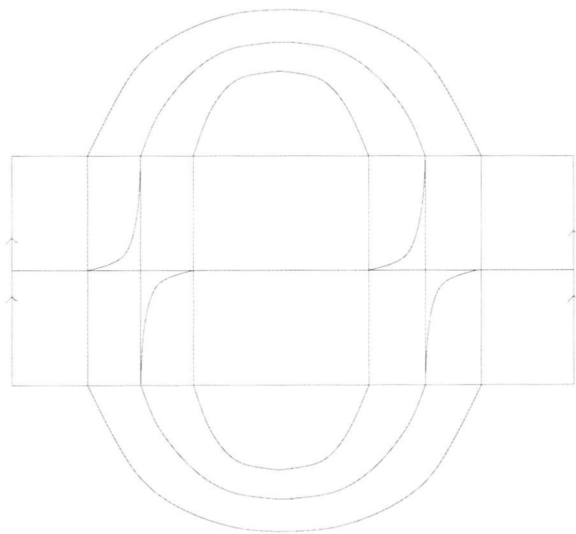

FIGURE 8. In the case $N = \mathbb{R}P^2$: The fiber M at $b = -1/2$

Now use the fact that we can trivialize $E_{r,s} \cap \{z \in E_{r,s} : r/2 < k(z) \leq r\}$ (where $k(z) = \frac{1}{4}(|z|^4 - |q_1(z)|^2)$ as in §2.6) by radial parallel transport, yielding an exact symplectomorphism

$$\rho : E_{r,s} \cap \{r/2 < k \leq r\} \longrightarrow \phi_1(D_{(r/2,r]}(T^*S^1)) \times D_s(c_1),$$

where $D_{(r/2,r]}(T^*S^1) = \{(u,v) \in T^*S^1 : |v| \in (r/2,r]\}$. We define

$$E_1 = F_0 \cup_\rho E_{r,s}$$

where we glue F_0 and $E_{r,s}$ using the map ρ. The map $\pi_1 : E_1 \longrightarrow D_s(c_1)$ is defined to be $\pi_{r,s} = z_1^2 - z_2^2 + c_1$ on $E_{r,s}$ and the projection map to $D_s(c_1)$ on F_0 (these two maps agree on the overlap using the gluing map ρ).

Now, the Lefschetz fibration $\pi : E \longrightarrow D^2$ is constructed such that the restriction of π to $\pi^{-1}(D_s(c_1))$ agrees with (E_1, π_1). Let

$$N_{r,s} = E_{r,s} \cap \mathbb{R}^2 \subset \mathbb{C}^2.$$

Set

$$f_1(x_1, x_2) = x_1^2 - x_2^2 + c_1,$$

so f_1 is the standard Morse function of index 1 on $\mathbb{R}^2$. Note that

$$\pi_{r,s}|N_{r,s} = f_1, \quad \text{and}$$

$$N_{r,s} = \{x \in \mathbb{R}^2 : |f_1(x)| \leq s, \ |x|^4 - f_1(x)^2 \leq r\}.$$

Note that $N_{r,s}$ is diffeomorphic to an 8 sided polygon as in figure 9 (i.e., $N_{r,s}$ is the same as N_1^{loc} in figure 9).

When we construct the embedding $N \subset E$ (see §6.1 below), we will write N as the union of several over-lapping pieces. These pieces correspond to something like a handle-decomposition of N. $N_{r,s}$ will play the role of the 1-handle. Let

$$I_+ = N_{r,s} \cap \{f_1 = s\} \quad \text{and} \quad I_- = N_{r,s} \cap \{f_1 = -s\}.$$

Then $I_\pm \cong S^0 \times D^1$. I_- corresponds to the part of the 1-handle which attaches to the 0-handle, and I_+ corresponds to the part of the 1-handle which meets the boundary of the 2-handle. We identify I_- with $S^0 \times D^1$ explicitly using:

$$\psi_- : S^0 \times [-a, a] \longrightarrow I_-,$$

$$\psi_-(\pm 1, \theta) = (\pm \sqrt{s} \sinh(\theta), \pm \sqrt{s} \cosh(\theta)),$$

where $a > 0$ is chosen suitably.

Using an explicit formula for ρ it is easy to check (see the proof of Lemma 7.2 in [J09]) that under the trivialization

$$\rho : E_{r,s} \cap \{r/2 < k \leq r\} \longrightarrow \phi_1(D_{(r/2,r]}(T^*S^1)) \times D_s(c_1),$$

we have $\rho(I_- \cap \{r/2 < k \leq r\})$ is equal to

$$D_{(r/2,r]}(T_a^* S^1) \cup D_{(r/2,r]}(T_b^* S^1)$$

for some $K_- = \{a, b\} \subset S^1$, $K_- \cong S^0$. We assume that ϕ_1 is chosen such that $\phi_1(K_-)$ is $V_0 \cap V_1$ in M and $\phi_1(D_r(T_a^* S^1) \cup D_r(T_b^* S^1))$ is equal to a neighborhood of $V_0 \cap V_1$ in V_0. The main technical ingredient for our construction of $N \subset E$ is the following lemma.

Lemma 2. Let $0 < a_0 < a$ be the unique number such that

$$\psi_-(S^0 \times ([-a, -a_0) \cup (a_0, a])) = I_- \cap \{r/2 < k \leq r\}.$$

Let ψ_-^0 denote the restriction $\psi_-|_{(S^0 \times ([-a, -a_0) \cup (a_0, a]))}$. Then under the trivialization

$$\rho : E_{r,s} \cap \{r/2 < k \leq r\} \longrightarrow \phi_1(D_{(r/2,r]}(T^* S^1)) \times D_s(c_1),$$

we have that

$$\phi_1 \circ \rho \circ \psi_-^0 : S^0 \times ([-a, -a_0) \cup (a_0, a]) \longrightarrow V_0$$

agrees with the framing that is used to attach the 1-handle in the given handle decomposition of N. More precisely, if $h : S^0 \times [-a, a] \longrightarrow \partial h_0$ is the attaching map of the 1-handle in the given handle decomposition of N, then the restriction $h|(S^0 \times ([-a, -a_0] \cup [a_0, a]))$ is equal to $\phi_1 \circ \rho \circ \psi_-^0$, up to isotopy.

The proof involves inspecting each map, and we find that rules (1) and (2), which we used for constructing M in §4.1, are precisely what we need to get this result.

6.1. Sketch of the exact Lagrangian embedding $N \subset E$

We describe $N \subset E$ as the union of several overlapping pieces. These correspond to something like a handle-decomposition of N. In fact, this type of decomposition is used in Milnor's book on the h-cobordism theorem [M65, pages 27-32]. For $N = \mathbb{R}P^2$ our decomposition will have four pieces as in figure 9 below. We will call this a Milnor type handle decomposition.

In the Milnor type handle decomposition of $\mathbb{R}P^2$ shown in figure 9 we have $N_0 = D^2$ and $N_2 = D^2$, which are the same as the usual 0- and 2-handles. Then there is

$$N_1^{loc} = \{x \in \mathbb{R}^2 : |f_1(x)| \leq \delta, |x|^4 - f_1(x)^2 \leq \epsilon\},$$

where $\delta, \epsilon > 0$ are some small numbers and $f_1(x) = x_1^2 - x_2^2$. Here, N_1^{loc} plays the role of the 1-handle, but it is diffeomorphic to polygon with eight edges (as opposed to a standard 1-handle, which is diffeomorphic to $D^1 \times D^1$). For the last piece, suppose that the $1-$handle (in the usual handle-decomposition) is attached using an embedding

$$\phi : S^0 \times [-\epsilon, \epsilon] \longrightarrow S^1 = \partial N_0.$$

Then the last piece is

$$N_1^{triv} = [S^1 \setminus \phi(S^0 \times (-\epsilon/2, \epsilon/2))] \times [-1, 1].$$

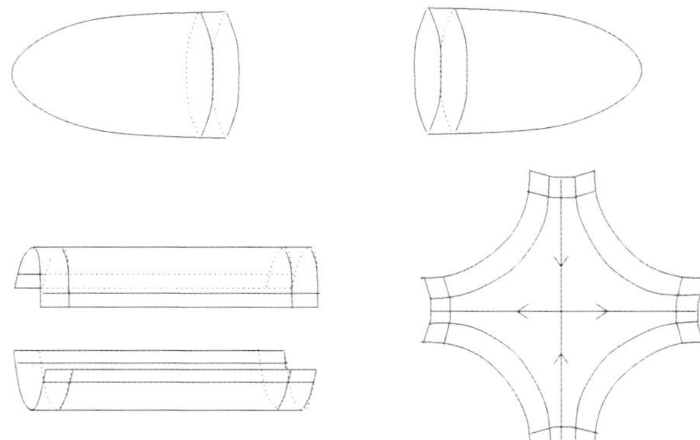

FIGURE 9. In the case $N = \mathbb{R}P^2$, the pieces N_0, N_2 (top), N_1^{triv} (bottom left), N_1^{loc} (bottom right). The overlap regions are also indicated.

This last piece has no analogue in a usual handle-decomposition; roughly, it fills in the rest of the space in N after N_0, N_2, N_1^{loc} are glued together. Now we realize four embeddings

$$(*) \quad N_0, N_1^{loc}, N_1^{triv}, N_2 \subset E$$

such that the four pieces overlap just like they do in the handle decomposition of $N = \mathbb{R}P^2$ (as in figure 9). Fix some small $\delta > 0$. Here is how we define the four embeddings $(*)$:

(1) N_0 is defined to be the Lefschetz thimble $\Delta_{[-3/4, -\delta]}$ with respect to the vanishing path $[-3/4, -\delta]$.

(2) N_2 is defined to be the Lefschetz thimble $\Delta_{[\delta, 3/4]}$ with respect to the vanishing path $[\delta, 3/4]$.

(3) To define N_1^{loc}, first take a neighborhood U of the critical point x_1, which lies over the middle critical value c_1. By construction of π (see above) we may assume $U = E_{\epsilon, 2\delta}$ (so $r = \epsilon, s = 2\delta$), and

$$\pi | E_{\epsilon, 2\delta} = \pi_{\epsilon, 2\delta}, \quad \text{where } \pi_{\epsilon, 2\delta}(z_1, z_2) = z_1^2 - z_2^2 + c_1.$$

Set

$$N_1^{loc} = N_{\epsilon, 2\delta} := E_{\epsilon, 2\delta} \cap \mathbb{R}^2, \quad \text{and}$$

$$f_1 = \pi_{\epsilon, 2\delta} | N_1^{loc}, \quad f_1(x_1, x_2) = x_1^2 - x_2^2 + c_1.$$

Explicitly,

$$N_1^{loc} = \{x \in \mathbb{R}^2 : |f_1(x)| \leq 2\delta, \ |x|^4 - f_1(x)^2 \leq \epsilon\}$$

for some small $\delta, \epsilon > 0$.

(4) We define N_1^{triv}. Consider $V_0 \subset M$ and let $\widetilde{V}_0$ denote the result of deleting a small neighborhood of $V_1 \cap V_0$ from V_0. Thus $\widetilde{V}_0$ is diffeomorphic to the disjoint union of two closed intervals: $\widetilde{V}_0 \cong [a, b] \cup [c, d]$. Now let N_1^{triv} be the result of parallel

transporting $\tilde{V}_0$ over the interval $[-3/4+\sigma, 3/4-\sigma]$, where $\sigma > 0$ is small enough that $b = -1/2$ and $b' = 1/2$ lie in the interior $(-3/4+\sigma, 3/4-\sigma)$. Obviously,

$$N_1^{triv} \cong ([a,b] \cup [c,d]) \times [-3/4+\sigma, 3/4-\sigma].$$

Note that the critical point x_1 does not cause a singularity to arise, since we deleted V_1 from V_0 to form $\tilde{V}_0$.

As we discussed just before Lemma 2, $f_1^{-1}(-\delta) \cap N_1^{loc}$ corresponds precisely to a neighborhood of $V_0 \cap V_1$ in V_0 (where V_0 here is understood to be transported from $M = \pi^{-1}(-1/2)$ to $\pi^{-1}(-\delta)$ along $[-1/2, -\delta]$). Moreover, Lemma 2 asserts that the embedding

$$f_1^{-1}(-\delta) \cap N_1^{loc} \longrightarrow V_0$$

precisely *agrees* with the framing used to attach h_1 to h_0 in the handle decomposition of $\mathbb{R}P^2$.

Similarly, there is an embedding

$$f_1^{-1}(\delta) \cap N_1^{loc} \longrightarrow V_0' \subset \pi^{-1}(\delta).$$

So, summarizing we can say there is are embeddings

$$f_1^{-1}(-\delta) \cap N_1^{loc} \longrightarrow \partial N_0$$

$$f_1^{-1}(\delta) \cap N_1^{loc} \longrightarrow \partial N_2$$

and the first one has the correct framing (agreeing with the one used to attach h_1 to h_0 in the standard handle decomposition of $\mathbb{R}P^2$). Now, N_1^{triv} simply fills in the gap from ∂N_0 to ∂N_1. We should define it so that it overlaps with N_0, N_2 and N_1^{loc}.

To conclude, the union $N_0 \cup N_1^{loc} \cup N_1^{triv} \cup N_2$ is diffeomorphic to N because it reproduces the Milnor style handle decomposition of N; the key point being that the correct framing is used for N_1^{loc} as it overlaps N_0. Moreover, N is an exact Lagrangian submanifold because for each piece θ restricts to be zero.

Remark 6.2. In the general case, i.e., if there are several critical points of index 1, then we have the same three vanishing paths, with $V_1^1, \ldots, V_1^k$ all having the same vanishing path $\gamma_1 = \gamma_1^j$, $j = 1, \ldots, k$. This is not a problem since all V_1^j are mutually disjoint. We make the same argument locally near each critical point of index 1. Of course N_1^{triv} is diffeomorphic to $(S^1 \setminus J) \times [a,b]$, where J is the union of several copies of $S^0 \times D^1$, disjointly embedded in $V_0 = S^1$.

6.2. Sketch proof of part 2 of Theorem

Part 2 of Theorem 6.1 essentially follows from inspection of the construction of $N \subset E$ given above. For instance $\pi(N_0) = [-3/4, -\delta]$ and $\pi|N_0$ obviously coincides (up to smooth reparameterization) with the usual model for an index 0 critical point on $\mathbb{R}^2$ given by the model $f_0 = -x_1^2 - x_2^2$.

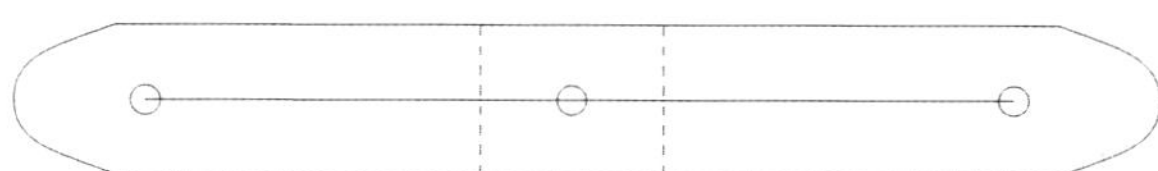

FIGURE 10. Here we have $A \subset \mathbb{C}$, a small neighborhood of $\pi(N) \subset \mathbb{R}$. Also A_0 is the region between the two dotted lines; A_- is the part of A to the left of A_0, and A_+ is to the right of A_0.

6.3. Sketch proof of $E \cong D(T^*N)$

First, use radial parallel transport to retract E onto $\pi^{-1}(A)$, where A is a small tubular neighborhood of $\pi(N) = [-3/4, 3/4] \subset D^2$. Let's split A into three pieces:

$$
\begin{aligned}
A_- &= A \cap \{x + iy : x < -\delta\} \\
A_+ &= A \cap \{x + iy : x > \delta\} \\
A_0 &= A \cap \{x + iy : -2\delta < x < 2\delta\}.
\end{aligned}
$$

See figure 10. Now let $E_- = \pi^{-1}(A_-)$. We construct (E, π) such that E_- consists of a trivial piece

$$
F_- = (M \setminus D_{\epsilon/2}(T^*V_0)) \times A_-
$$

with a copy of the local model with fiber $D_\epsilon(T^*V_0)$ glued on. Now consider the subset $E_-' \subset E_-$ given by

$$
E_-' = (M \setminus D_\epsilon(T^*V_0)) \times A_- \subset F_-.
$$

Now, let

$$
\begin{aligned}
A_0^- &= A \cap \{x + iy : -2\delta < x < -\delta\} \subset A_0, \\
A_0^+ &= A \cap \{x + iy : \delta < x < 2\delta\} \subset A_0.
\end{aligned}
$$

And set

$$
E_0^\pm = \pi^{-1}(A_0^\pm).
$$

Now we transport almost the whole region E_-' into $E_0^- = \pi^{-1}(A_0^-)$ along paths parallel to the real line, so that the points corresponding to $M \setminus D_\epsilon(T^*V_0)$ move onto the corresponding points in the fibers of $E_0 = \pi^{-1}(A_0^-)$. See figure 11 for a schematic picture.

We say we move "almost" all of E_-' because what we do more precisely is the following. Fix some small fixed $a > 0$. Let $p = (p', z) \in E_-' = (M \setminus D_\epsilon(T^*V_0)) \times A_-$. If $p' \in M \setminus D_\epsilon(T^*V_0)$ has distance $> a$ from $D_\epsilon(T^*V_0)$ then we transport p, over a path parallel to the real line, into $E_0^- = \pi^{-1}(A_0^-)$ (for a time depending only on $Re(z)$, where $z = \pi(p) \in A_-$). But if $p' \in M \setminus D_\epsilon(T^*V_0)$ has distance $t \in [0, a]$ from $D_\epsilon(T^*V_0)$ then we multiply the old transport time by a cut-off function $\phi(t)$, so as to taper down to the identity map. This means not quite all of E_-' is transported out of E_- into E_0. But the part that remains can be retracted in the fiber direction (by a Liouville flow) onto the local model over A_- with fiber $D_\epsilon(T^*V_0)$.

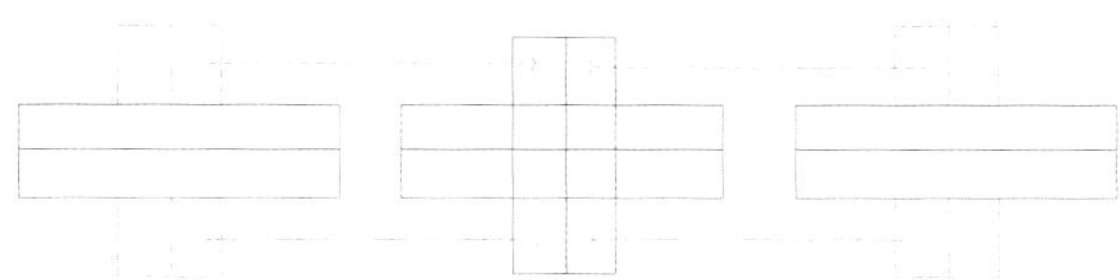

FIGURE 11. Here we have a schematic drawing of the retraction map. The three crosses represent the regular fibers over A_-, A_0, and A_+. The dotted parts of the crosses on the left and right correspond to the part of the fiber corresponding to $M \setminus D(T^*V_0)$. The idea of the retraction is to parallel transport those dotted parts of the fibers over A_- and A_+ onto the corresponding parts of the fibers over A_0.

We do a similar procedure to E_+, transporting (most) of a set $E_+' \subset E_+$ into E_0^+, and then retracting what remains in the fiber direction onto the local model over A_+ with fiber $D_\epsilon(T^*V_2')$. (Here, (E, π) is constructed such that the part over A_+ consists of a trivial piece

$$F_+ = (M' \setminus D_{\epsilon/2}(T^*V_2')) \times A_+$$

with a copy of the local model with fiber $D_\epsilon(T^*V_2')$ glued on.)

So what we have reduced to so far in E_+ and E_- is an arbitrarily small neighborhood of N_0 and N_2. Indeed, in the base direction we began by retracting onto a small neighborhood of $\pi(N) = [-3/4, 3/4]$ and in the fiber direction, we are inside the local models with fiber $D_\epsilon(T^*V_0)$ over A_- and with fiber $D_\epsilon(T^*V_2)$ over A_+. Thus, in E_+ and E_- we have retracted (by a Liouville type flow) onto small neighborhoods of N_0 and N_1 which can be modeled by $D(T^*N_0)$ and $D(T^*N_2)$.

To finish our discussion, let's take a look at E_0. As in the discussion before Lemma 2, (E, π) is constructed such that E_0 consists of a trivial piece

$$F_0 = (M \setminus D_{\epsilon/2}(T^*V_1)) \times A_0$$

with a copy of the local model with fiber $D_\epsilon(T^*V_1)$ glued on.

The part of E_0 given by the local model over A_0 with fiber $D_\epsilon(T^*V_1)$ is a neighborhood of N_1^{loc} in $\mathbb{C}^2$ which is isomorphic to $D(T^*N_1^{loc})$. Next, note that N_1^{triv} is given by

$$N_1^{triv} = (V_0 \setminus D_{\epsilon/2}(T^*V_1)) \times A_0 \subset F_0.$$

The trivial part of E_0 given by F_0 can be retracted to an arbitrarily small neighborhood N_1^{triv} as follows. First we deal with the fiber direction by shrinking the radius of $D(T^*V_0) \subset M$ enough (by a fiberwise retraction). Then, in the base direction we make A_0 close enough to $\pi(N_1^{triv}) = [-2\delta, 2\delta]$. In this way F_0 retracts onto a small neighborhood of N_1^{triv} which can be modeled by $D(T^*N_1^{triv})$.

To conclude we have retracted E (using a Liouville type flow) onto a union of neighborhoods of the form $D(T^*N_0)$, $D(T^*N_2)$, $D(T^*N_1^{triv})$, $D(T^*N_1^{loc})$, which together form a Weinstein neighborhood of N in E, isomorphic to $D(T^*N)$.

References

[AS10]   M. Abouzaid, and P. Seidel, Altering symplectic manifolds by homologous recombination, arXiv preprint, arXiv:1007.3281.

[AC99]   N. A'Campo, Real deformations and complex topology of plane curve singularities, *Ann. Fac. Sci. Toulouse Math. (6)* **8** (1999), 5–23.

[AGZV88]   V. I. Arnold, S. M. Gusein-Zade, and A. N. Varchenko, *Singularities of differentiable maps. Vol. II. Monodromy and asymptotics of integrals*, Monographs in Mathematics, vol. 83, Birkhauser, 1988.

[AMP05]   D. Auroux, V. Muñoz, and F. Presas, Lagrangian submanifolds and Lefschetz pencils, *J. Symplectic Geom.*, **3** (2005), no. 2, 171–219.

[AS04]   D. Auroux and I. Smith, Lefschetz pencils, branched covers and symplectic invariants, in Lect. Notes in Math., no. 1938, Springer, 2008, 1–53.

[FSS08]   K. Fukaya, P. Seidel, and I. Smith, Exact Lagrangian submanifolds in simply-connected cotangent bundles, *Invent. Math.*, **172** (2008), no. 1, 1–27.

[GS99]   R. Gompf and A. Stipsicz, *4-Manifolds and Kirby Calculus*, Graduate Studies in Math. Vol. 20, AMS, 1999.

[J09]   J. Johns, The Picard-Lefschetz theory of Complexifications of Morse functions, arXiv preprint, arXiv:0906.1218

[L81]   K. Lamotke, The topology of complex projective varieties after S. Lefschetz., *Topology*, **20** (1981), no. 1, 15–51.

[M09]   M. Maydanskiy, Exotic symplectic manifolds from Lefschetz fibrations, arXiv preprint, arXiv:0906.2224.

[MS09]   M. Maydanskiy and P. Seidel, Lefschetz fibrations and exotic symplectic structures on cotangent bundles of spheres, *J. Topol.*, **3** (2010), no. 1, 157-180.

[M65]   J. Milnor, *Lectures on the h-cobordism theorem*, Princeton Mathematical Notes, Princeton University Press, 1965.

[P91]   L. Polterovich, The surgery of Lagrange submanifolds, *Geom. Funct. Anal.*, **1** (1991), no. 2, 198–210.

[S03A]   P. Seidel, A long exact sequence for Floer cohomology, *Topology*, **42** (2003), no. 5, 1003–1063.

[S03B]   P. Seidel, Homological mirror symmetry for the quartic surface, arXiv preprint, arXiv:math.SG/0310414.

[S06]   P. Seidel, Lectures on four-dimensional Dehn twists, in Symplectic 4-manifolds and algebraic surfaces, 231–267, Lecture Notes in Math., no. 1938, Springer, Berlin, 2008

[S08]   P. Seidel, *Fukaya categories and Picard-Lefschetz theory*, Zurich Lectures in Advanced Mathematics, EMS, 2008.

[W91]   A. Weinstein, Contact surgery and symplectic handlebodies, *Hokkaido Math. J.*, **20** (1991), no. 2, 241–251.

MAX PLANCK INSTITUTE, INSELSTR. 22, LEIPZIG 04103, GERMANY
E-mail address: johns@mis.mpg.de

Nine-dimensional exceptional quotient singularities exist

Ivan Cheltsov and Constantin Shramov

ABSTRACT. We prove that nine-dimensional exceptional quotient singularities exist.

The Fubini–Studi metric on $\mathbb{P}^n$ is known to be Kähler–Einstein. Moreover, it follows from [1] that every Kähler–Einstein metric on $\mathbb{P}^n$ is a pull back of the Fubini–Studi metric (possibly multiplied by a positive real constant) via some automorphism of $\mathbb{P}^n$. However, there are plenty of non-Kähler–Einstein metrics on $\mathbb{P}^n$ whose Kähler forms lie in $c_1(\mathbb{P}^n)$. Let $g = g_{i\bar{j}}$ be such a metric with a Kähler form ω. Then one can try to obtain the Kähler–Einstein metric on $\mathbb{P}^n$ out of the metric g by taking the normalized Kähler–Ricci iterations defined by

$$\begin{cases} \omega_{i-1} = \mathrm{Ric}\big(\omega_i\big), \\ \omega_0 = \omega, \end{cases} \tag{1}$$

where ω_i is a Kähler form such that $\omega_i \in c_1(\mathbb{P}^n)$. Indeed, it follows from [19] that the solution ω_i to (1) exists for every $i \geqslant 1$. However, it is not clear that any solution to (1) converges to the Kähler form of the Kähler–Einstein metric on $\mathbb{P}^n$ in the sense of Cheeger–Gromov (see [15]). Nevertheless, this is known to be true under an additional assumption that we are going to describe.

Let $\bar{G} \subset \mathrm{Aut}(\mathbb{P}^n)$ be a finite subgroup. Suppose, in addition, that the metric g is $\bar{G}$-invariant. Let $\alpha_{\bar{G}}(\mathbb{P}^n)$ be the $\bar{G}$-invariant α-invariant of Tian of $\mathbb{P}^n$ that is introduced in [18].

Theorem 2 ([15]). *If $\alpha_{\bar{G}}(\mathbb{P}^n) > 1$, then any solution to (1) converges to the Kähler form of the Kähler–Einstein metric on $\mathbb{P}^n$ in $C^\infty(X)$-topology.*

It should be mentioned that the original result by Rubinstein proved in [15] is much stronger than Theorem 2 and is valid for any smooth complex manifold with a positive first Chern class. Nevertheless, even in the simplest possible case of $\mathbb{P}^n$, the assertion of Theorem 2 is still very not obvious. Thus, it is natural to ask the following

Question 3 (Rubinstein). *Is there a finite subgroup $\bar{G} \subset \mathrm{Aut}(\mathbb{P}^n)$ such that $\alpha_{\bar{G}}(\mathbb{P}^n) > 1$?*

It came as a surprise that Question 3 is strongly related to the notion of exceptional singularity that was introduced by Shokurov in [16]. Let us recall this notion.

We assume that all varieties are projective, normal, and defined over $\mathbb{C}$.

Definition 4 (Shokurov). *Let $(V \ni O)$ be a germ of Kawamata log terminal singularity. Then $(V \ni O)$ is said to be* exceptional *if for every effective $\mathbb{Q}$-divisor D_V on the variety V such that (V, D_V) is log canonical and for every resolution of singularities $\pi\colon U \to V$ there exists at most one π-exceptional divisor $E \subset U$ such that $a(V, D_V, E) = -1$, where the rational number $a(V, D_V, E)$ can be defined through the equivalence*

$$K_U + D_U \sim_{\mathbb{Q}} \pi^*\!\left(K_V + D_V\right) + \sum a\!\left(V, D_V, F\right) F.$$

The sum above is taken over all f-exceptional divisors, and D_U is the proper transform of the divisor D_V on the variety U.

One can show that exceptional Kawamata log terminal singularities are straightforward generalizations of the Du Val singularities of type $\mathbb{E}_6$, $\mathbb{E}_7$ and $\mathbb{E}_8$ (see [16, Example 5.2.3]), which partially justifies the word "exceptional" in Definition 4. It follows from our earlier result [3, Theorem 1.16] that exceptional Kawamata log terminal singularities exist in every dimension. Surprisingly, Question 3 is *almost* equivalent to the following

Question 5. *Are there exceptional* quotient *singularities of dimension $n + 1$?*

Recall that quotient singularities are always Kawamata log terminal. So Question 5 fits well to Definition 4. It follows from [16], [13], [3], and [4] that the answers to both Questions 3 and 5 are positive for every $n \leqslant 5$ (see Theorems 10 and 12). Moreover, it follows from [4] that the answers to both Questions 3 and 5 are "surprisingly" negative for $n = 6$. The purpose of this paper is to show that the answers to both Questions 3 and 5 are again positive for $n = 8$ by proving the following

Theorem 6. *Let G be a finite subgroup in $\mathrm{SL}_9(\mathbb{C})$ such that $G \cong 3^{1+4} : \mathrm{Sp}_4(3)$ (see [6] for notation), let $\phi\colon \mathrm{SL}_9(\mathbb{C}) \to \mathrm{Aut}(\mathbb{P}^8)$ be the natural projection. Put $\bar{G} = \phi(G)$. Then $4/3 \geqslant \alpha_{\bar{G}}(\mathbb{P}^8) \geqslant 10/9$ and the singularity $\mathbb{C}^9/G$ is exceptional.*

How to compute $\alpha_{\bar{G}}(\mathbb{P}^n)$? How to show that a given quotient singularity is exceptional? How are Questions 3 and 5 related? How to prove Theorem 6? What are the expected answers to Questions 3 and 5 for $n = 7$ and $n \geqslant 9$? Let us give partial answers to these questions.

Let $\phi\colon \mathrm{GL}_{n+1}(\mathbb{C}) \to \mathrm{Aut}(\mathbb{P}^n)$ be the natural projection. Then there exists a finite subgroup G in $\mathrm{GL}_{n+1}(\mathbb{C})$ such that $\phi(G) = \bar{G}$. Put

$$\mathrm{lct}\!\left(\mathbb{P}^n, \bar{G}\right) = \sup \left\{ \lambda \in \mathbb{Q} \ \middle| \ \begin{array}{l} \text{the log pair } (\mathbb{P}^n, \lambda D) \text{ has log canonical singularities} \\ \text{for every } \bar{G}\text{-invariant effective } \mathbb{Q}\text{-divisor } D \sim_{\mathbb{Q}} -K_{\mathbb{P}^n} \end{array} \right\}.$$

Theorem 7 (see e.g., [2, Theorem A.3]). *One has $\mathrm{lct}(\mathbb{P}^n, \bar{G}) = \alpha_{\bar{G}}(\mathbb{P}^n)$.*

The number $\mathrm{lct}(\mathbb{P}^n, \bar{G})$ is usually called $\bar{G}$-equivariant *global log canonical threshold* of $\mathbb{P}^n$. Despite the fact that $\mathrm{lct}(\mathbb{P}^n, \bar{G}) = \alpha_{\bar{G}}(\mathbb{P}^n)$, we still prefer to work with the number $\mathrm{lct}(\mathbb{P}^n, \bar{G})$ throughout this paper, because it is easier to handle than $\alpha_{\bar{G}}(\mathbb{P}^n)$. For example, it follows immediately from the definition of the number $\mathrm{lct}(\mathbb{P}^n, \bar{G})$ that $\mathrm{lct}(\mathbb{P}^n, \bar{G})$ is less than or equal to $d/(n+1)$ if the group G has a semi-invariant of degree d (a semi-invariant of the group G is a polynomial whose zeroes define a $\bar{G}$-invariant hypersurface in $\mathbb{P}^n$).

Recall that an element $g \in G$ is called a *reflection* (or sometimes a *quasi-reflection*) if there is a hyperplane in $\mathbb{P}^n$ that is pointwise fixed by $\phi(g)$. To answer Question 5 one can always assume that the group G does not contain reflections (cf. [4, Remark 1.16]). On the other hand, one can easily check that there exists a finite subgroup $G' \subset \mathrm{SL}_{n+1}(\mathbb{C})$ such that $\phi(G') = \bar{G}$. So to answer Question 3 one can also assume that $G \subset \mathrm{SL}_{n+1}(\mathbb{C})$, which implies, in particular, that the group G does not contain reflections. Moreover, if the group G does not contain reflections, then the singularity $\mathbb{C}^{n+1}/G$ is exceptional if and only if the singularity $\mathbb{C}^{n+1}/G'$ is exceptional thanks to the following

Theorem 8 ([3, Theorem 3.17]). *Let G be a finite subgroup in $\mathrm{GL}_{n+1}(\mathbb{C})$ that does not contain reflections. Then the singularity $\mathbb{C}^{n+1}/G$ is exceptional if and only if for any $\bar{G}$-invariant effective $\mathbb{Q}$-divisor D on $\mathbb{P}^n$ such that $D \sim_{\mathbb{Q}} -K_{\mathbb{P}^n}$, the log pair $(\mathbb{P}^n, D)$ is Kawamata log terminal.*

Corollary 9. *Let G be a finite subgroup in $\mathrm{GL}_{n+1}(\mathbb{C})$ that does not contain reflections. Then*

- *the singularity $\mathbb{C}^{n+1}/G$ is exceptional if $\mathrm{lct}(\mathbb{P}^n, \bar{G}) > 1$,*
- *the singularity $\mathbb{C}^{n+1}/G$ is not exceptional if either $\mathrm{lct}(\mathbb{P}^n, \bar{G}) < 1$,*
- *the singularity $\mathbb{C}^{n+1}/G$ is not exceptional if G has a semi-invariant of degree at most $n + 1$,*
- *for any subgroup $G' \subset \mathrm{GL}_{n+1}(\mathbb{C})$ such that G' does not contain reflections and $\phi(G') = \bar{G}$, the singularity $\mathbb{C}^{n+1}/G$ is exceptional if and only if the singularity $\mathbb{C}^{n+1}/G'$ is exceptional.*

The assumption that G does not contain reflections is crucial for Theorem 8 (see [3, Example 1.18]). On the other hand, it follows from [14, Proposition 2.1] that G must be primitive (see for example [3, Definition 1.21]) if $\mathbb{C}^{n+1}/G$ is exceptional. Moreover, for small $n \leqslant 4$, we have the following

Theorem 10 ([13, Theorem 1.2], [3, Theorem 1.22]). *Let G be a finite subgroup in $\mathrm{GL}_{n+1}(\mathbb{C})$ that does not contain reflections. Suppose that $n \leqslant 4$. Then the following conditions are equivalent:*

- *the singularity $\mathbb{C}^{n+1}/G$ is exceptional,*
- *$\mathrm{lct}(\mathbb{P}^n, \bar{G}) \geqslant (n + 2)/(n + 1)$,*
- *the group G is primitive and has no semi-invariants of degree at most $n + 1$.*

In particular, both Questions 3 and 5 are equivalent for $n \leqslant 4$ and can be expressed in terms of primitivity and absence of semi-invariants of small degree of the group G. It appears that in higher dimensions the latter is no longer true, since there are non-exceptional six-dimensional quotient singularities arising from primitive subgroups without reflections in $\mathrm{GL}_6(\mathbb{C})$ that have no semi-invariants of degree at most 6 (see [3, Example 3.25]). On the other hand, we still believe that both Questions 3 and 5 are equivalent for every n, which can be summarized as

Conjecture 11. *Let G be a finite subgroup in $\mathrm{GL}_{n+1}(\mathbb{C})$ that does not contain reflections. Then the singularity $\mathbb{C}^{n+1}/G$ is exceptional if and only if $\mathrm{lct}(\mathbb{P}^n, \bar{G}) > 1$.*

In fact, Conjecture 11 still holds for $n = 5$, because of

Theorem 12 ([4, Theorem 1.14]). *Let G be a finite subgroup in $\mathrm{SL}_6(\mathbb{C})$. Then the following are equivalent:*

- *the singularity $\mathbb{C}^6/G$ is exceptional,*
- *the inequality $\mathrm{lct}(\mathbb{P}^5, \bar{G}) \geqslant 7/6$ holds,*
- *either $\bar{G}$ is the Hall–Janko sporadic simple group (see [12]), or $G \cong 6.\mathrm{A}_7$ and $\bar{G} \cong \mathrm{A}_7$.*

In particular, both Questions 3 and 5 are equivalent and both have positive answers for $n = 5$. For $n = 6$, both Questions 3 and 5 are also equivalent and both have negative answers due to

Theorem 13 ([4, Theorem 1.16]). *For every finite subgroup G in $\mathrm{GL}_7(\mathbb{C})$, the singularity $\mathbb{C}^7/G$ is not exceptional and $\mathrm{lct}(\mathbb{P}^n, \bar{G}) \leqslant 1$.*

To apply Theorem 8 we may assume that $G \subset \mathrm{SL}_{n+1}(\mathbb{C})$, since there exists a finite subgroup $G' \subset \mathrm{SL}_{n+1}(\mathbb{C})$ such that $\phi(G') = \bar{G}$. On the other hand, it is well known that there are at most finitely many primitive finite subgroups in $\mathrm{SL}_{n+1}(\mathbb{C})$ up to conjugation by Jordan's theorem for complex linear groups. Primitive finite subgroups of $\mathrm{SL}_2(\mathbb{C})$ are group-theoretic counterparts of Platonic solids and each of them gives rise to an exceptional quotient singularity (see [16, Example 5.2.3]). Similar classification is possible in small dimensions. For example, primitive finite subgroups of $\mathrm{SL}_{n+1}(\mathbb{C})$ for $n \leqslant 6$ have been classified long time ago (see for example [7]). This allowed to obtain the complete list of all finite subgroups in $\mathrm{SL}_{n+1}(\mathbb{C})$ for every $n \leqslant 6$ that give rise to exceptional quotient singularities (see [13], [3], and [4]), which implies, in particular, that the answers to both Questions 3 and 5 are positive for every $n \leqslant 5$ and are negative for $n = 6$ (see Theorem 13). We have no idea right now what are the answers to Questions 3 and 5 in the cases when $n = 7$ and $n \geqslant 9$, but we expect that the answers to both Questions 3 and 5 may still be negative for all $n \gg 0$ due to the following

Theorem 14. *Let G be the finite subgroup in $\mathrm{GL}_{n+1}(\mathbb{C})$ such that $\bar{G}$ is a sporadic simple group. Then $\mathbb{C}^{n+1}/G$ is exceptional if and only if $n = 5$ and $\bar{G}$ is the Hall–Janko sporadic simple group.*

Proof. Since $\bar{G}$ is simple, we may assume that G has no quasi-reflections. Explicit computations in GAP (see [8]) imply that G has a semi-invariant of degree at most $n+1$ (and thus $\mathbb{C}^{n+1}/G$ is not exceptional by Theorem 8) unless $n = 5$ and $\bar{G}$ is the Hall–Janko sporadic simple group[1]. If $n = 5$ and $\bar{G}$ is the Hall–Janko sporadic simple group, then the singularity $\mathbb{C}^{n+1}/G$ is exceptional by [4, Theorem 1.14]. $\qquad\square$

---

[1]Similarly, one can show that G has a semi-invariant of degree at most n (and thus $\mathbb{C}^{n+1}/G$ is not weakly-exceptional (see [3, Definition 3.7]) by [3, Theorem 3.16]) unless either $n = 5$ and $\bar{G}$ is the Hall–Janko sporadic simple group, or $n = 11$ and $\bar{G}$ is the Suzuki sporadic simple group (see [17]). We expect that in the latter case the corresponding quotient singularity is actually weakly-exceptional.

An indirect evidence that both Questions 3 and 5 may have negative answers for all $n \gg 0$ is given by

Theorem 15 ([5]). *Let G be the finite primitive subgroup in $\mathrm{GL}_{n+1}(\mathbb{C})$. Suppose that $n \geqslant 12$. Then $|\bar{G}| \leqslant (n+2)!$. Moreover, if $|\bar{G}| = (n+2)!$, then $\bar{G} \cong \mathrm{S}_{n+2}$.*

In fact, Collins obtained the optimal bounds for $|\bar{G}|$ for every $n \leqslant 11$ if G is primitive (his proof uses known lower bounds for the degrees of the faithful representations of each quasisimple group, for which the classification of finite simple groups is required). Moreover, it follows from [16], [13], [5], [3], and [4] that $|\bar{G}|$ reaches its maximum on a subgroup $\bar{G}$ in $\mathrm{Aut}(\mathbb{P}^n)$ with $\mathrm{lct}(\mathbb{P}^n, \bar{G}) > 1$ if $n \leqslant 3$, and this is no longer true for $4 \leqslant n \leqslant 6$. Surprisingly, it follows from Theorem 6 that in the case when $n = 8$, the number $|\bar{G}|$ reaches its maximum if G is isoclinic to a finite subgroup in $\mathrm{GL}_9(\mathbb{C})$ that is mentioned in Theorem 6. For $n = 11$, the number $|\bar{G}|$ reaches its maximum if $\bar{G}$ is the Suzuki sporadic simple group.

In the remaining part of the paper we prove Theorem 6. Let G be a finite subgroup in $\mathrm{SL}_9(\mathbb{C})$ from Theorem 6. Then the embedding $G \hookrightarrow \mathrm{SL}_9(\mathbb{C})$ is given by an irreducible nine-dimensional G-representation[2], which we denote by U.

The outline of the proof of Theorem 6 is as follows. We assume that $\mathrm{lct}(\mathbb{P}^8, \bar{G}) < 10/9$ and seek for a contradiction. There exists a $\bar{G}$-invariant $\mathbb{Q}$-divisor D on $\mathbb{P}^8$ and a positive rational number $\lambda < 10/9$ such that $D \sim_{\mathbb{Q}} -K_{\mathbb{P}^8}$ and the log pair $(\mathbb{P}^8, \lambda D)$ is strictly log canonical, i.e., log canonical and not Kawamata log terminal. Arguing as in [3] and [4], we apply Nadel–Shokurov vanishing (see [11, Theorem 9.4.8]) and Kawamata subadjunction (see [10, Theorem 1]) to obtain restrictions on the Hilbert polynomial of the minimal center of log canonical singularities of the log pair $(\mathbb{P}^8, \lambda D)$ (see [9, Definition 1.3], [10]). Composing the latter with results coming from representation theory we obtain a contradiction. One of the few new ingredients of the proof is the *binomial trick* (see Lemma 27).

Let us list without proofs some properties of the G-representation U (Lemmas 16, 17, 18, and 19) that can be verified by direct computations. We used GAP (see [8]) to carry them out.

Lemma 16. *The group G does not have semi-invariants of degree $d \leqslant 11$, and there exists a semi-invariant of the group G of degree 12.*

Denote by Δ_k the collection of dimensions of irreducible subrepresentations of $\mathrm{Sym}^k(U^\vee)$. We will use the following notation: writing $\Delta_k = [\ldots, r \times m, \ldots]$, we mean that among the irreducible subrepresentations of $\mathrm{Sym}^k(U^\vee)$ there are exactly r subrepresentations of dimension m (not necessarily isomorphic to each other). Furthermore, denote by Σ_k the set of partial sums of Δ_k, i.e., the set of all numbers $s = \sum r'_i m_i$, where $\Delta_k = [\ldots, r_1 \times m_1, \ldots, r_2 \times m_2, \ldots, r_i \times m_i, \ldots]$ and $0 \leqslant r'_i \leqslant r_i$ for all i. We use the abbreviation m_i for $1 \times m_i$.

---

[2]Note that the group G has two irreducible representations of dimension 9, but they differ only by an outer automorphism of G, so that the *subgroup* $G \subset \mathrm{SL}_9(\mathbb{C})$ is defined uniquely up to conjugation.

Lemma 17. *The representation* $\mathrm{Sym}^2(U^\vee)$ *is irreducible (and has dimension 45). Futhermore,* $\Delta_3 = [5, 160]$, $\Delta_4 = [45, 180, 270]$, $\Delta_5 = [36, 90, 135, 216, 270, 540]$,

$$\Delta_6 = [4, 15, 24, 2 \times 80, 3 \times 240, 3 \times 480, 640],$$

$$\Delta_7 = [9, 36, 3 \times 135, 3 \times 180, 3 \times 216, 270, 324, 405, 3 \times 540, 720, 2 \times 729],$$

$$\Delta_8 = [36, 4 \times 45, 5 \times 180, 5 \times 270, 2 \times 324, 6 \times 360, 3 \times 405, 7 \times 540, 2 \times 576, 720, 729],$$

$$\Delta_9 = [3 \times 5, 3 \times 20, 3 \times 30, 40, 45, 60, 80, 12 \times 160, 12 \times 240, 10 \times 480, 10 \times 640, 11 \times 720].$$

Lemma 18. *Let* $U_{45} \subset \mathrm{Sym}^4(U^\vee)$ *be the 45-dimensional irreducible subrepresentation. Then*

$$U^\vee \otimes U_{45} \cong U_{90} \oplus U_{135} \oplus U_{180}$$

as a G-representation, where U_{90}, U_{135} *and* U_{180} *are irreducible G-representations of dimensions 90, 135 and 180, respectively.*

Lemma 19. *Let* $U_{24} \subset \mathrm{Sym}^6(U^\vee)$ *be the 24-dimensional irreducible subrepresentation. Then* $U^\vee \otimes U_{24}$ *is an irreducible G-representation.*

Now we are ready to prove Theorem 6. It follows from Theorems 7 and 8 that to prove Theorem 6 it is enough to prove that $4/3 \geqslant \mathrm{lct}(\mathbb{P}^8, \bar{G}) \geqslant 10/9$. On the other hand, one has $\mathrm{lct}(\mathbb{P}^8, \bar{G}) \leqslant 4/3$ by Lemma 16. In fact, we believe that $\mathrm{lct}(\mathbb{P}^8, \bar{G}) = 4/3$, but we are unable to prove this now. To complete the proof of Theorem 6, we must prove that $\mathrm{lct}(\mathbb{P}^8, \bar{G}) \geqslant 10/9$. Suppose that $\mathrm{lct}(\mathbb{P}^8, \bar{G}) < 10/9$. Then there is an effective $\bar{G}$-invariant $\mathbb{Q}$-divisor $D \sim_\mathbb{Q} \mathcal{O}_{\mathbb{P}^8}(9)$, and there is a positive rational number $\lambda < 10/9$ such that $(\mathbb{P}^8, \lambda D)$ is strictly log canonical.

Let S be a minimal center of log canonical singularities of the log pair $(\mathbb{P}^8, \lambda D)$ (see [9, Definition 1.3], [10]), let V be the $\bar{G}$-orbit of the subvariety $S \subset \mathbb{P}^8$, and let r be the number of irreducible components of the subvariety V. Then $\deg(V) = r \cdot \deg(S)$.

Arguing as in the proofs of [9, Theorem 1.10] and [10, Theorem 1], we may assume that the only log canonical centers of the log pair $(\mathbb{P}^8, \lambda D)$ are components of the subvariety V (see [3, Lemma 2.8]). Then it follows from [9, Proposition 1.5] that the components of the subvariety V are disjoint. In particular, if $\dim(V) \geqslant 4$, then $r = 1$. Put $n = \dim(V)$ which also equals $\dim(S)$. Then $n \neq 7$ by Lemma 16.

Let $\mathcal{I}_V$ be the ideal sheaf of the subvariety $V \subset \mathbb{P}^8$, and let Λ be a general hyperplane in $\mathbb{P}^8$. Put $H = \Lambda|_V$, $h_m = h^0(\mathcal{O}_V(mH))$ and $q_m = h^0(\mathcal{O}_{\mathbb{P}^8}(m) \otimes \mathcal{I}_V)$ for every $m \in \mathbb{Z}$. It follows from the Shokurov–Nadel vanishing theorem (see [11, Theorem 9.4.8]) that

$$q_m = h^0\left(\mathcal{O}_{\mathbb{P}^8}(m)\right) - h_m = \binom{8+m}{m} - h_m$$

for every $m \geqslant 1$. In particular, if $n = 0$, then $r = h_1 \leqslant 9$, which is impossible by Lemma 16. Hence, we see that $1 \leqslant n \leqslant 6$.

It follows from [10, Theorem 1] that the variety V is normal and has at most rational singularities. Moreover, it follows from [10, Theorem 1] that for every positive rational number $\epsilon > 0$ there is an effective $\mathbb{Q}$-divisor B_V on the variety V such that

$$\left(K_{\mathbb{P}^n} + \lambda D + \epsilon \Lambda\right)\Big|_V \sim_{\mathbb{Q}} K_V + B_V,$$

and (V, B_V) has Kawamata log terminal singularities. In particular, taking ϵ sufficiently small, we may assume that $K_V + B_V \sim_{\mathbb{Q}} (1 - \nu)H$ for some positive $\nu \in \mathbb{Q}$, because $\lambda < 10/9$.

Remark 20. One can show that any irreducible representation W of the group G such that the center $Z(G) \cong \mathbb{Z}_3$ acts non-trivially on W has dimension $\dim(W)$ divisible by 9. Therefore one has $h_i \equiv 0 \mod 9$ for every i not divisible by 3, since $Z(G)$ acts nontrivially on $\mathrm{Sym}^i(U^\vee)$ if i is not divisible by 3.

Remark 21. One has $q_1 = 0$ since $U^\vee$ is irreducible and $q_2 = 0$ by Lemma 17.

Put $d = H^n = \deg(V)$ and $H_V(m) = \chi(\mathcal{O}_V(m))$. Then the Shokurov–Nadel vanishing theorem (see [11, Theorem 9.4.8]) implies that $H_V(m) = h_m$ for every $m \geqslant 1$. Recall that $H_V(m)$ is a Hilbert polynomial of the subvariety V, which is a polynomial in m of degree n with leading coefficient $d/n!$.

Lemma 22. *For any non-negative integer δ one has*

$$d = h_{\delta+n+1} - \binom{n}{1} h_{\delta+n} + \binom{n}{2} h_{\delta+n-1} + \ldots + (-1)^n h_{\delta+1}.$$

Proof. Induction by n. $\qquad\square$

Lemma 23. *If $4 \leqslant n \leqslant 5$, then d is divisible by 3. If $n = 6$, then d is divisible by 9.*

Proof. Suppose that $n = 6$. Applying Lemma 22 with $\delta = 0$ and $\delta = 1$, we get

$$h_7 - 6h_6 + 15h_5 - 20h_4 + 15h_3 - 6h_2 + h_1 = d = h_8 - 6h_7 + 15h_6 - 20h_5 + 15h_4 - 6h_3 + h_2, \quad (24)$$

which gives $21h_6 - 21h_3 \equiv 0 \mod 9$ by subtracting two equalities in (24), reducing everything modulo 9, and using Remark 20. Thus $h_6 - h_3 \equiv 0 \mod 3$, and $15h_6 - 6h_3 \equiv 0 \mod 9$. The latter equality combined with (24) implies that $d \equiv 0 \mod 9$.

If $n = 5$, a similar argument shows that $d \equiv 0 \mod 3$.

Finally, suppose that $n = 4$. Applying Lemma 22 for $\delta = 0$, we get

$$h_5 - 4h_4 + 6h_3 - 4h_2 + h_1 = d,$$

which gives $d \equiv 6h_3 \equiv 0 \mod 3$, since h_5, h_4, h_2, and h_1 are divisible by 3 by Remark 20. $\qquad\square$

Remark 25. If $n = 6$, then $q_3 = 0$. Indeed, if $q_3 > 0$, then $q_3 > 2$ by Lemma 17, so that $d < 9$, which is impossible by Lemma 23. Similarly, if $n = 6$ and $q_4 > 0$, one has $d = 9$.

Remark 26. One has $h_m \leqslant h_{m+1}$ and $q_m \leqslant q_{m+1}$ for all $m \geqslant 1$.

Let $\Lambda_1, \Lambda_2, \ldots, \Lambda_n$ be general hyperplanes in $\mathbb{P}^8$. Put $\Pi_j = \Lambda_1 \cap \ldots \cap \Lambda_j$, $V_j = V \cap \Pi_j$, $H_j = V_j \cap H$, and $B_{V_j} = B_V|_{V_j}$ for every $j \in \{1, \ldots, n\}$. Put $V_0 = V$, $B_{V_0} = B_V$, $H_0 = H$, $\Pi_0 = \mathbb{P}^8$. For every $j \in \{0, 1, \ldots, n\}$, let $\mathcal{I}_{V_j}$ be the ideal sheaf of the subvariety $V_j \subset \Pi_j$. Recall that $\Pi_j \cong \mathbb{P}^{8-j}$ and put $q_i(V_j) = h^0(\mathcal{O}_{\Pi_j}(i) \otimes \mathcal{I}_{V_j})$ for every $j \in \{0, 1, \ldots, n\}$.

Lemma 27. *Suppose that $i \geqslant j + 1$ and $j \in \{1, \ldots, n\}$. Then*

$$q_i(V_j) = q_i - \binom{j}{1} q_{i-1} + \binom{j}{2} q_{i-2} - \ldots + (-1)^j q_{i-j}.$$

Proof. For every $j \in \{0, 1, \ldots, n\}$, it follows from the adjunction formula that

$$K_{V_j} + B_{V_j} \sim_{\mathbb{Q}} (j + 1 - \nu) H_j,$$

because $K_V + B_V \sim_{\mathbb{Q}} (1 - \nu)H$ and (V_j, B_{V_j}) has at most Kawamata log terminal singularities. Applying the Nadel–Shokurov vanishing theorem to the log pair (V_j, B_{V_j}), we see that $h^1(\mathcal{O}_{V_j}(i)) = 0$ for every $i \geqslant j + 1$ and every $j \in \{0, 1, \ldots, n\}$. Thus, we have

$$h^0\left(\mathcal{O}_{V_j}\left((i+1)H_j\right)\right) - h^0\left(\mathcal{O}_{V_j}\left(iH_j\right)\right) = h^0\left(\mathcal{O}_{V_{j+1}}\left((i+1)H_{j+1}\right)\right) \tag{28}$$

for every $i \geqslant j + 1$ and every $j \in \{0, \ldots, n - 1\}$. Now applying the Nadel–Shokurov vanishing theorem to the log pair $(\Pi_j, \lambda D|_{\Pi_j})$, we see that $h^1(\mathcal{O}_{\Pi_j}(i) \otimes \mathcal{I}_{V_j}) = 0$ for every $i \geqslant j + 1$ and every $j \in \{0, 1, \ldots, n\}$. This implies that

$$q_i(V_j) = \binom{8 - j + i}{i} - h^0\left(\mathcal{O}_{V_j}\left(iH_j\right)\right) \tag{29}$$

for every $i \geqslant j + 1$ and every $j \in \{0, 1, \ldots, n\}$. Combining (28) and (29), we have
$q_i(V_{j-1}) - q_{i-1}(V_{j-1})$

$$= \binom{9 - j}{i} - h^0\left(\mathcal{O}_{V_{j-1}}\left(iH_{j-1}\right)\right) - \binom{8 - j + i}{i - 1} + h^0\left(\mathcal{O}_{V_{j-1}}\left((i - 1)H_{j-1}\right)\right) =$$

$$= \binom{9 - j}{i} - \binom{8 - j + i}{i - 1} - h^0\left(\mathcal{O}_{V_j}\left(iH_j\right)\right) = \binom{8 - j + i}{i} - h^0\left(\mathcal{O}_{V_j}\left(iH_j\right)\right) = q_i(V_j)$$

for every $i \geqslant j + 1$ and every $j \in \{1, \ldots, n\}$. Thus, we see that

$$q_i(V_j) = q_i(V_{j-1}) - q_{i-1}(V_{j-1}) \tag{30}$$

for every $i \geqslant j + 1$ and every $j \in \{1, \ldots, n\}$. Iterating (30), we obtain the required equality. $\square$

Lemma 27 allows one to obtain bounds on the numbers q_i.

Remark 31. There are trivial bounds $0 \leqslant q_i(V_j) < \binom{8-j+i}{i}$.

Recall that $q_1 = q_2 = 0$ by Remark 21, and $q_3 = 0$ if $n = 6$ by Remark 25. Therefore, Remark 31 implies

Corollary 32. *If $n = 5$, one has*

$$0 \leqslant q_4 - 3q_3 < 126. \tag{33}$$

If $n = 6$, one has

$$0 \leqslant q_5 - 4q_4 < 126. \tag{34}$$

Playing with the numbers $q_i(V_j)$, we can obtain

Lemma 35. *Suppose that $n \geqslant 4$. Then*

$$\binom{9}{n} - \frac{nd}{2} > q_n(V_{n-1}) \geqslant \binom{9}{n} - nd - 1.$$

Proof. Recall that the variety $V_{n-1} \subset \mathbb{P}^{8-n+1}$ is a smooth curve of degree d, since V is normal. Since $n \geqslant 4$, we see that V_{n-1} is irreducible. Let g be the genus of the curve V_{n-1}. It follows from the adjunction formula that $K_{V_{n-1}} + B_{V_{n-1}} \sim_{\mathbb{Q}} (n - \nu)H_{n-1}$, because $K_V + B_V \sim_{\mathbb{Q}} (1 - \nu)H$. In particular, one has $2g - 2 < dn$.

Applying the Nadel–Shokurov vanishing theorem to the log pair $(\Pi_{n-1}, \lambda D|_{\Pi_{n-1}})$, we see that

$$q_m(V_{n-1}) = \binom{8 - n + 1 + m}{m} - h^0\left(\mathcal{O}_{V_{n-1}}\left(mH_{n-1}\right)\right)$$

for every $m \geqslant n$. Since $2g - 2 < dn$, the divisor nH_{n-1} is non-special. Therefore, it follows from the Riemann–Roch theorem that

$$q_n(V_{n-1}) = \binom{9}{n} - nd + g - 1,$$

which implies the required inequalities, since $2g - 2 < dn$ and $g \geqslant 0$. $\square$

Combining Lemmas 35 and 27 and recalling the trivial bounds from Remark 31, we obtain

Corollary 36. *If $n = 4$, then*

$$\max\left(0, 125 - 4d\right) \leqslant q_4 - 3q_3 \leqslant 125 - 2d. \tag{37}$$

If $n = 5$, then

$$\max\left(0, 125 - 5d\right) \leqslant q_5 - 4q_4 + 6q_3 \leqslant 126 - \frac{5d}{2}. \tag{38}$$

If $n = 6$, then

$$0 \leqslant q_6 - 5q_5 + 10q_4 - 10q_3 \leqslant 83 - 3d. \tag{39}$$

As a by-product of Corollary 36, we get

Corollary 40. *If $n = 4$, then $d \leqslant 62$. If $n = 5$, then $d \leqslant 50$. If $n = 6$, then $d \leqslant 27$.*

The above restrictions reduce the problem to a combinatorial question of finding all polynomials H_V of degree n with a leading coefficient $d/n!$, such that $h_m = H_V(m) \in \Sigma_m$ for sufficiently many $m \geqslant 1$, and such that the numbers h_m and $q_m = h^0(\mathcal{O}_{\mathbb{P}^8}(m)) - h_m$ satisfy the conditions arising from Lemma 23, Corollaries 32, 36 and 40, and Remarks 21, 25 and 26. This can be done in a straighforward way, although the number of cases to be considered is so large that we had to delegate this part of the proof to a simple computer program. Finally, we get the following four lemmas which we leave without proofs.

Lemma 41. *There are no polynomials* $H(m)$ *of degree* $n \leqslant 3$ *such that the values* $h_m = H(m)$ *are in* Σ_m *for* $1 \leqslant m \leqslant 6$ *and* $h_i \leqslant h_{i+1}$ *for* $1 \leqslant i \leqslant 5$.

Lemma 42. *If* $H(m)$ *is a polynomial of degree* $n = 4$ *with a leading coefficient* $d/n!$ *with* $d \leqslant 62$ *and* d *divisible by* 3, *such that the values* $h_m = H(m)$ *are in* Σ_m *for* $1 \leqslant m \leqslant 6$, *and the numbers* h_m *and* $q_m = \binom{8+m}{m} - h_m$ *satisfy the bounds of Remark 26 and* (37), *then* $d = 36$, $q_1 = q_2 = q_3 = 0$, $q_4 = 45$, $q_5 = 270$.

Lemma 43. *If* $H(m)$ *is a polynomial of degree* $n = 5$ *with a leading coefficient* $d/n!$ *with* $d \leqslant 50$ *and* d *divisible by* 3, *such that the values* $h_m = H(m)$ *are in* Σ_m *for* $1 \leqslant m \leqslant 9$, *and the numbers* h_m *and* $q_m = \binom{8+m}{m} - h_m$ *satisfy the bounds of Remark 26,* (38) *and* (33), *then* $d = 45$, $q_1 = q_2 = q_3 = q_4 = q_5 = 0$, $q_6 = 39$, $q_7 = 270$.

Lemma 44. *There are no polynomials* $H(m)$ *of degree* $n = 6$ *with a leading coefficient* $d/n!$ *with* $d \leqslant 27$ *and* d *divisible by* 9, *such that the values* $h_m = H(m)$ *are in* Σ_m *for* $1 \leqslant m \leqslant 9$, *and the numbers* h_m *and* $q_m = \binom{8+m}{m} - h_m$ *satisfy the bounds of Remark 26,* (39) *and* (34).

Applying Lemmas 41, 42, 43, and 44 to the Hilbert polynomial $H_V(m)$, we end up with the following two possibilities: either $n = 4$, $d = 36$, $q_1 = q_2 = q_3 = 0$, $q_4 = 45$, $q_5 = 270$, or $n = 5$, $d = 45$, $q_1 = q_2 = q_3 = q_4 = q_5 = 0$, $q_6 = 39$, $q_7 = 270$.

Remark 45. Let W be a G-subrepresentation in $H^0(\mathcal{I}_V \otimes \mathcal{O}_{\mathbb{P}^8}(m))$. Then there is a natural map of G-representations $\psi \colon U^\vee \otimes W \to H^0(\mathcal{I}_V \otimes \mathcal{O}_{\mathbb{P}^8}(m+1))$, which is obviously a non-zero map.

Let us suppose that $n = 4$. Then $q_4 = 45$, so that by Lemma 17 there is an irreducible 45-dimensional G-subrepresentation $U_{45} \subset H^0(\mathcal{I}_V \otimes \mathcal{O}_{\mathbb{P}^8}(4))$. Thus, there is a morphism of G-representations $\psi_5 \colon U^\vee \otimes U_{45} \to H^0(\mathcal{I}_V \otimes \mathcal{O}_{\mathbb{P}^8}(5))$, which is a non-zero map by Remark 45. On the other hand, $H^0(\mathcal{I}_V \otimes \mathcal{O}_{\mathbb{P}^8}(5))$ has no 180-dimensional G-subrepresentations by Lemma 17. Put $q_5' = \dim(\mathrm{Im}\psi_5)$. Keeping in mind the splitting of $U^\vee \otimes U_{45}$ described in Lemma 18, we see that q_5' equals either 90, or 135, or 225. Since $q_5 = 270$, there must exist a (possibly reducible) G-subrepresentation of $H^0(\mathcal{I}_V \otimes \mathcal{O}_{\mathbb{P}^8}(5))$ of dimension $q_5 - q_5'$, i.e., of dimension 180, 135 and 45, respectively. Neither of these cases is possible by Lemma 17 (in particular, the second case is impossible since $H^0(\mathcal{I}_V \otimes \mathcal{O}_{\mathbb{P}^8}(5))$ contains a unique G-invariant subspace of dimension 135). Therefore, one has $n \neq 4$.

Finally, we see that $n = 5$. Then $q_6 = 39$, so that by Lemma 17 there is an irreducible 24-dimensional G-subrepresentation $U_{24} \subset H^0(\mathcal{I}_V \otimes \mathcal{O}_{\mathbb{P}^8}(6))$. Therefore there is a morphism of G-representations $\psi_7 \colon U^\vee \otimes U_{24} \to H^0(\mathcal{I}_V \otimes \mathcal{O}_{\mathbb{P}^8}(7))$, which is a non-zero map by Remark 45. Since $U^\vee \otimes U_{24}$ is an irreducible 216-dimensional G-representation by Lemma 19, we see that ψ_7 is injective. Since $q_7 = 270$, there must exist a (possibly reducible) G-subrepresentation of $H^0(\mathcal{I}_V \otimes \mathcal{O}_{\mathbb{P}^8}(7))$ of dimension $270 - 216 = 54$, which is impossible by Lemma 17. The obtained contradiction completes the proof of Theorem 6.

Acknowledgements: This work was partially supported by the grants N.Sh.-4713.2010.1, RFFI 11-01-00336-a. RFFI 11-01-92613-KO-a, RFFI 08-01-00395-a, RFFI 11-01-00185-a, LMS grant 41007 and NSF DMS-1001427.

We would like to thank Michael Collins for a reference to Theorem 15. Special thanks goes to Andrey Zavarnitsyn who explained us various facts from Group Theory and provided us with the results of GAP computations used in the proof of Theorem 14. The first author would like to thank personally Selman Akbulut for inviting him to Eighteenth Gökova Geometry and Topology conference (Gökova, Turkey).

References

[1] S. Bando and T. Mabuchi, Uniqueness of Einstein Kahler metrics modulo connected group actions, *Advanced Studies in Pure Mathematics*, **10** (1987), 11–40.

[2] I. Cheltsov and C. Shramov, Log canonical thresholds of smooth Fano threefolds, *Russian Mathematical Surveys*, **63** (2008), 73–180.

[3] I. Cheltsov and C. Shramov, On exceptional quotient singularities, *Geometry and Topology*, **15** (2011), 1843–1882.

[4] I. Cheltsov and C Shramov, Six-dimensional exceptional quotient singularities, Mathematical Research Letters, to appear.

[5] M. Collins, Bounds for finite primitive complex linear groups, *J. of Algebra*, **319** (2008), 759–776.

[6] J. Conway, R. Curtis, S. Norton, R. Parker, and R. Wilson, *Atlas of finite groups*, Clarendon Press, Oxford, 1985.

[7] W Feit, The current situation in the theory of finite simple groups, in Actes du Congrès International des Mathématiciens, Gauthier–Villars, Paris (1971), 55–93.

[8] The GAP Group *GAP – Groups, Algorithms, and Programming, Version 4.4.12*, available at http://www.gap-system.org.

[9] Y. Kawamata, On Fujita's freeness conjecture for 3-folds and 4-folds, *Mathematische Annalen*, **308** (1997), 491–505.

[10] Y. Kawamata, Subadjunction of log canonical divisors II, *American Journal of Mathematics*, **120** (1998), 893–899.

[11] R. Lazarsfeld, *Positivity in algebraic geometry II* Springer-Verlag, Berlin, 2004.

[12] J. Lindsey, On a six dimensional projective representation of the Hall–Janko group, *Pacific Journal of Mathematics*, **35** (1970), 175–186.

[13] D. Markushevich and Yu. Prokhorov, Exceptional quotient singularities, *American Journal of Mathematics*, **121** (1999), 1179–1189.

[14] Y Prokhorov, Sparseness of exceptional quotient singularities, *Mathematical Notes*, **68** (2000), 664–667.

[15] Y. Rubinstein, Some discretizations of geometric evolution equations and the Ricci iteration on the space of Kähler metrics, *Advances in Mathematics*, **218** (2008), 1526–1565.

[16] V. Shokurov, Three-fold log flips, *Izvestiya Mathematics*, Russian Academy of Sciences, **40** (1993), 95–202.

[17] M. Suzuki, A simple group of order 448,345,497,600, in Symposium "Theory of Finite Groups" (1968), Benjamin, New York, 1969, 113–119.

[18] G. Tian, On Kähler–Einstein metrics on certain Kähler manifolds with $c_1(M) > 0$, *Inventiones Mathematicae*, **89** (1987), 225–246.

[19] S. T. Yau, On the Ricci curvature of a compact Kähler manifold and the complex Monge–Ampère equation, I, *Communications on Pure and Applied Mathematics*, **31** (1978), 339–411.

UNIVERSITY OF EDINBURGH, EDINBURGH EH9 3JZ, UK, I.CHELTSOV@ED.AC.UK

STEKLOV INSTITUTE OF MATHEMATICS, MOSCOW 119991, RUSSIA, SHRAMOV@MCCME.RU

LABORATORY OF ALGEBRAIC GEOMETRY, GU-HSE, 7 VAVILOVA STREET, MOSCOW 117312, RUSSIA

Proceedings of 18th Gŏkova
Geometry-Topology Conference
pp. 97 – 124

Landau–Ginzburg models — old and new

Ludmil Katzarkov and Victor Przyjalkowski

ABSTRACT. In the last three years a new concept — the concept of wall crossing has emerged. The current situation with wall crossing phenomena, after papers of Seiberg–Witten, Gaiotto–Moore–Neitzke, Vafa–Cecoti and seminal works by Donaldson–Thomas, Joyce–Song, Maulik–Nekrasov–Okounkov–Pandharipande, Douglas, Bridgeland, and Kontsevich–Soibelman, is very similar to the situation with Higgs Bundles after the works of Higgs and Hitchin — it is clear that a general "Hodge type" of theory exists and needs to be developed. Nonabelian Hodge theory did lead to strong mathematical applications —uniformization, Langlands program to mention a few. In the wall crossing it is also clear that some "Hodge type" of theory exists — Stability Hodge Structure (SHS). This theory needs to be developed in order to reap some mathematical benefits — solve long standing problems in algebraic geometry. In this paper we look at SHS from the perspective of Landau–Ginzburg models and we look at some applications. We consider simple examples and explain some conjectures these examples suggest.

1. Introduction

Mirror symmetry is a physical duality between $N = 2$ superconformal field theories. In the 1990's Maxim Kontsevich reinterpreted this concept from physics as an incredibly deep and far-reaching mathematical duality now known as Homological Mirror Symmetry (HMS). In a famous lecture in 1994, he created a frenzy in the mathematical community which lead to synergies between diverse mathematical disciplines: symplectic geometry, algebraic geometry, and category theory. HMS is now the cornerstone of an immense field of active mathematical research.

In the last three years a new concept — the concept of wall crossing has emerged. The current situation with wall crossing phenomena, after papers of Seiberg–Witten, Gaiotto–Moore–Neitzke, Vafa–Cecoti and seminal works by Donaldson–Thomas, Joyce–Song, Maulik–Nekrasov–Okounkov–Pandharipande, Douglas, Bridgeland, and Kontsevich–Soibelman, is very similar to the situation with Higgs Bundles after the works of Higgs and Hitchin — it is clear that a general "Hodge type" of theory exists and needs to be developed. Nonabelian Hodge theory did lead to strong mathematical applications

Key words and phrases. Hodge structures; categories; Landau–Ginzburg models.

L. K was funded by NSF Grant DMS0600800, NSF FRG Grant DMS-0652633, FWF Grant P20778, and an ERC Grant — GEMIS, V. P. was funded by FWF grant P20778, RFFI grants 11-01-00336-a and 11-01-00185-a, grants MK−1192.2012.1, NSh−5139.2012.1, and AG Laboratory GU-HSE, RF government grant, ag. 11 11.G34.31.0023.

— uniformization, Langlands program to mention a few. In the wall crossing it is also clear that some "Hodge type" of theory needs to be developed in order to reap some mathematical benefits — solve long standing problems in algebraic geometry.

The foundations of these new Hodge structures, which we call *Stability Hodge Structures (SHS)* will appear in a paper by the first author, Kontsevich, Pantev and Soibelman (see [1]). In this paper we will look at SHS from the perspective of Landau–Ginzburg models and we will also look at some applications. We will consider simple examples and explain some conjectures these examples suggest. Further elaboration and examples will appear in [1] and [2].

We start with the classical interpretation of wall crossings in Landau–Ginzburg models. After that we describe a hypothetical program of "Stability Hodge Theory" which combines Nonabelian and Noncommutative Hodge theory. We consider some possible applications in this paper. First we consider an approach to the conjecture that the universal covering of a smooth projective variety is holomorphically convex. This is a classical question in algebraic geometry proven by the first author and collaborators for linear fundamental groups [3]. It was believed that for nonresidually finite fundamental groups one needs a different approach and in this paper we outline a procedure of extending the argument to the nonresidually finite case based on SHS. We also outline possible applications to Hodge structures with many filtrations and to Sarkisov's theory.

Stability Hodge Structure is a notion which originates from functions of one complex variable and combinatorics — gaps, polygons, and circuits. We give these classical notions a new read through HMS and category theory, dressing them up with some cluster varieties and integrable systems. After that we enhance these data additionally with some basic nonabelian Hodge theory in order to get a property we need — strictness. In the same way as moduli spaces of Higgs bundles parameterize spectral coverings, the moduli space of deformed stability conditions parameterizes Landau–Ginzburg models.

We believe this is only the tip of the iceberg and this very rich motivic conglomerate of ideas will play an important role in the studies of categories and of algebraic cycles. In particular we suggest that the categorical notion of spectra can be seen as a Hodge theoretic notion related to the "homotopy type of a category".

The paper is organized as follows. In Sections 2 and 3 we describe the classical approach to Landau–Ginzburg models and wall crossings. After that in Sections 4, 5, 6 we define Stability Hodge Structures and build a parallel with Simpson's nonabelian Hodge theory. We also discuss possible applications in Sections 7, 8, 9.

2. Classical Landau–Ginzburg models and wall crossings

In this section we recall the "classical" way of interpreting wall crossing in the case of Landau–Ginzburg models. We will establish a certain combinatorial framework on which we later base our constructions.

We recall the notion of Landau–Ginzburg models from the Laurent polynomials point of view. For more details see, say, [4] and references therein.

Let X be a smooth Fano variety of dimension N. We can associate *a quantum co-homology ring* $QH^*(X) = H^*(X, \mathbb{Q}) \otimes \Lambda$ to it, where Λ is the Novikov ring for X. The multiplication in this ring, the so called *quantum multiplication*, is given by *(genus zero) Gromov–Witten invariants* — numbers counting rational curves lying in X. Given these data one can associate *a regularized quantum differential operator* Q_X (the second Dubrovin connection) — the regularization of an operator associated with connection in the trivial vector bundle given by a quantum multiplication by the canonical class K_X. In "good" cases such as we consider (for Fano threefolds or complete intersections) the equation $Q_X I = 0$ has a unique normalized analytic solution $I = 1 + a_1 t + a_2 t^2 + \ldots$.

Definition 2.1. *A toric Landau–Ginzburg model* is a Laurent polynomial $f \in \mathbb{C}[x_1^{\pm 1}, \ldots, x_n^{\pm}]$ such that:

> **Period condition:** The constant term of $f^i \in \mathbb{C}[x_1^{\pm 1}, \ldots, x_n^{\pm}]$ is a_i for any i (this means that I is a period of a family $f \colon (\mathbb{C}^*)^n \to \mathbb{C}$, see [4]).
>
> **Calabi–Yau condition:** Any fiber of $f \colon (\mathbb{C}^*)^n \to \mathbb{C}$ after some fiberwise compact-ification has trivial dualizing sheaf.
>
> **Toric condition:** There is an embedded degeneration $X \rightsquigarrow T$ to a toric variety T whose fan polytope (the convex hull of generators of its rays) coincides with the Newton polytope (the convex hull of non-zero coefficients) of f. A Laurent polynomial without the toric condition is called *a weak Landau–Ginzburg model*.

Toric Landau–Ginzburg models for complete intersections can be derived from the Hori–Vafa suggestions (see, say, [5]).

Definition 2.2. Let X be a general Fano complete intersection of hypersurfaces of degrees $d_1, \ldots, d_k$ in $\mathbb{P}^n$. Let $d_0 = n - d_1 - \ldots - d_k$ be its index. Then a Laurent polynomial

$$f_X = \frac{(x_{1,1} + \ldots + x_{1,d_1-1} + 1)^{d_1} \cdot \ldots \cdot (x_{k,1} + \ldots + x_{k,d_k-1} + 1)^{d_k}}{\prod x_{ij}} + x_{01} + \ldots + x_{0d_0-1}.$$

we call *of Hori–Vafa type*.

Theorem 2.3 (Proposition 9 in [5] and Theorem 2.2, [6]). *The polynomial f_X is a toric Landau–Ginzburg model for X.*

Definition 2.4. Let f be a Laurent polynomial in $\mathbb{C}[x_0^{\pm 1}, \ldots, x_n^{\pm 1}]$. Then a (non-toric birational) symplectomorphism is called *of cluster type* if it is a composition of toric change of variables and symplectomorphisms of type

$$y_0 = x_0 \cdot f_0(x_1, \ldots, x_i)^{\pm 1}, \quad y_1 = x_1, \ldots, y_n = x_n,$$

for some Laurent polynomial f_0 and under this change of variables f goes to a Laurent polynomial for which a Calabi–Yau condition holds.

It is called *elementary of cluster type* if (up to toric change of variables)

$$f_0 = x_1 + \ldots + x_i + 1.$$

It is called *of linear cluster type* if it is a composition of elementary symplectomorphisms of cluster type and toric change of variables.

Remark 2.5. For all examples in the rest of the paper the Calabi–Yau condition holds for all considered cluster type transformations.

Proposition 2.6. *Let f be a weak Landau–Ginzburg model for X. Let f' be a Laurent polynomial obtained from f by symplectomorphism of cluster type. Then f' is a weak Landau–Ginzburg model for X.*

Proof. A period giving constant terms of Laurent polynomials is, up to proportion, an integral of the (depending on $\lambda \in \mathbb{C}$) form $\frac{1}{1-\lambda f} \prod \frac{dx_i}{x_i}$ over a standard n-cycle on the torus $|x_1| = \ldots = |x_n| = 1$. This integral does not change under cluster type symplecto-morphisms. $\square$

Example 2.7. Let X be a quadric threefold. There are two types of degenerations of X to normal toric varieties inside the space of quadratic forms. That is,

$$T_0 = \{x_1 x_2 = x_2^3\} \subset \mathbb{P}[x_1 : x_2 : x_3 : x_4 : x_5]$$

and

$$T_1 = \{x_1 x_2 = x_3 x_4\} \subset \mathbb{P}[x_1 : x_2 : x_3 : x_4 : x_5].$$

Let

$$f_0 = \frac{(x+1)^2}{xyz} + y + z$$

be a weak Landau–Ginzburg model of Hori–Vafa type for X. Let

$$f_1 = \frac{(x+1)}{xyz} + y(x+1) + z$$

be its cluster-type transformation given by the change of variables

$$\frac{y}{(x+1)} \mapsto y.$$

One can see that $T_0 = T_{f_0}$ and $T_1 = T_{f_1}$.

Remark 2.8. One can see that applying the same change of variables a second time to f_1 gives back (up to toric change of variables) f_0.

Example 2.9. Let X be a cubic threefold. There are two types of degenerations of X to normal toric varieties inside the space of cubic forms. That is,

$$T_0 = \{x_1 x_2 x_3 = x_4^3\} \subset \mathbb{P}[x_1 : x_2 : x_3 : x_4 : x_5]$$

and

$$T_1 = \{x_1 x_2 x_3 = x_4^2 x_5\} \subset \mathbb{P}[x_1 : x_2 : x_3 : x_4 : x_5].$$

Let

$$f_0 = \frac{(x+y+1)^3}{xyz} + z$$

be a weak Landau–Ginzburg model of Hori–Vafa type for X. Let

$$f_1 = \frac{(x+y+1)^2}{xyz} + z(x+y+1)$$

be its cluster-type transformation given by the change of variables

$$\frac{z}{(x+y+1)} \mapsto z.$$

One can see that $T_0 = T_{f_0}$ and $T_1 = T_{f_1}$.

Remark 2.10. Applying this change of variables a second time to f_1 we get (up to toric change of variables) f_1 again and applying it a third time we get f_0 back.

Example 2.11. Let X be a cubic fourfold. There are three types of degenerations of X to normal toric varieties inside the space of cubic forms. That is,

$$T_{00} = \{x_1 x_2 x_3 = x_4^3\} \subset \mathbb{P}[x_1 : x_2 : x_3 : x_4 : x_5 : x_6],$$

$$T_{10} = \{x_1 x_2 x_3 = x_4^2 x_5\} \subset \mathbb{P}[x_1 : x_2 : x_3 : x_4 : x_5 : x_6],$$

and

$$T_{11} = \{x_1 x_2 x_3 = x_4 x_5 x_6\} \subset \mathbb{P}[x_1 : x_2 : x_3 : x_4 : x_5 : x_6],$$

Let

$$f_{00} = \frac{(x+y+1)^3}{xyzt} + z + t$$

be a weak Landau–Ginzburg model of Hori–Vafa type for X. Let

$$f_{10} = \frac{(x+y+1)^2}{xyzt} + z(x+y+1) + t$$

be its cluster-type transformation given by the change of variables

$$\frac{z}{(x+y+1)} \mapsto z$$

and let

$$f_{11} = \frac{(x+y+1)}{xyzt} + z(x+y+1) + t(x+y+1)$$

be the cluster-type transformation of f_{10} given by the change of variables

$$\frac{t}{(x+y+1)} \mapsto t.$$

One can see that $T_{00} = T_{f_{00}}$, $T_{10} = T_{f_{10}}$, and $T_{11} = T_{f_{11}}$.

Remark 2.12. Applying the first change of variables a second time to f_{10} we get (up to toric change of variables) f_{10} again, applying it once more we get f_{00}, and applying any change of variables to f_{11} we get f_{10}.

Example 2.13. Consider quadrics in $\mathbb{P} = \mathbb{P}(1,1,1,1,2)$. Denote the coordinates in $\mathbb{P}$ by x_0, x_1, x_2, x_3, x_4, where the weight of x_4 is 2. The general quadric is

$$T_1 = \{F_2(x_0, x_1, x_2, x_3) + \lambda x_4 = 0\},$$

where F_2 is a quadratic form and $\lambda \in \mathbb{C} \setminus 0$. Projection on the hyperplane generated by $x_0, \ldots, x_3$ gives an isomorphism of T_1 with $\mathbb{P}^3$. The general variety with $\lambda = 0$ is a toric variety

$$T_2 = \{x_0 x_1 = x_2 x_3\}.$$

It degenerates to

$$T_3 = \{x_1 x_2 = x_0^2\}.$$

One can see that T_3 is an image of $\mathbb{P}(1,1,2,4)$ under the Veronese map v_2.

Consider the following 3 weak Landau–Ginzburg models for $\mathbb{P}^3$:

$$f_1 = x + y + z + \frac{1}{xyz},$$

$$f_2 = x + \frac{y}{x} + \frac{z}{x} + \frac{1}{xy} + \frac{1}{xz},$$

$$f_3 = \frac{(x+1)^2}{xyz} + \frac{y}{z} + z.$$

Changing toric variables one can rewrite f_1 as

$$f_1' = z(x+1) + y + \frac{1}{xyz^2},$$

$$f_1'' = z(x+1) + \frac{y}{z} + \frac{1}{xyz}.$$

The cluster-type change of variables

$$x \mapsto x, \quad y \mapsto y, \quad z(x+1) \mapsto z$$

sends f_1' to a Laurent polynomial that differs from f_3 by a toric change of variables and f_1'' to a polynomial

$$z + \frac{(x+1)y}{z} + \frac{(x+1)}{xyz},$$

which differs from f_2 by toric change of variables. The cluster-type change of variables

$$x \mapsto x, \quad y(x+1) \mapsto y, \quad z \mapsto z$$

sends the last expression to f_3.

One can see that $T_1 = T_{f_1}$, $T_2 = T_{f_2}$, and $T_3 = T_{f_3}$.

Theorem 2.14 (Hacking–Prokhorov, [7]). *Let X be a degeneration of $\mathbb{P}^2$ to a $\mathbb{Q}$-Gorenstein surface with quotient singularities. Then $X = \mathbb{P}(a^2, b^2, c^2)$, where (a,b,c) is any solution of the Markov equation $a^2 + b^2 + c^2 = 3abc$.*

Remark 2.15. All Markov triples are obtained from the basic one $(1,1,1)$ by a sequence of *elementary transforms*

$$(a, b, c) \mapsto (a, b, 3ab - c).$$

Proposition 2.16 (S. Galkin). *Let (a, b, c) be a Markov triple and let f be a weak Landau–Ginzburg model for $\mathbb{P}^2$ such that $T_f = \mathbb{P}(a^2, b^2, c^2)$. Then there is an elementary cluster-type transformation such that for the image f' of f under this transformation $T_{f'} = \mathbb{P}(a^2, b^2, (3ab - c)^2)$.*

Sketch of the proof (S. Galkin). Consider $d \geq c$ such that $3ad = b \pmod{c}$. One can check that we can choose toric coordinates x, y such that in these coordinates vertices of the Newton polytope of f are (d, c), $(d - c, c)$, and $(-\frac{d(3ab-c)-b^2}{c}, -3ab + c)$. Let p be the k-th integral point from the end of an edge of integral length n of the Newton polytope of f. Then the coefficient of f at p is $\binom{n}{k}$ (this can be proved by induction). This means that

$$f = x^{d-c} y^c (x + 1)^c + \frac{1}{x^{\frac{d(3ab-c)-b^2}{c}} y^{3ab-c}} + \sum_r y^{-n_r} f_r(x),$$

where n_i's are non-negative and f_i's are some Laurent polynomials in x. One can check that the change of variables of cluster type

$$y' = y(x + 1), \quad x' = x$$

sends f to a weak Landau–Ginzburg model f' such that $T_{f'} = \mathbb{P}(a^2, b^2, (3ab - c)^2)$. $\quad\square$

We extend observed connection between degenerations and birational transformations further to a general connection between geometry of moduli space of Landau–Ginzburg models, birational and symplectic geometry. We summarize this connection in Table 1 and we will investigate it (mainly conjecturally) in the sections that follow.

	Fano variety X	Landau–Ginzburg model $LG(X)$
A side	$\mathsf{Fuk}(X)$: symplectomorphisms and general degenerations	$FS(LG(X))$: degenerations
B side	$D^b_{sing}(LG(X))$: phase changes	$D^b(X)$: birational transformations

Table 1. Wall crossings.

3. Minkowski decompositions and cluster transformations

Definition 3.1. Let $N \cong \mathbb{Z}^n$ be a lattice. Denote $N_\mathbb{R} = N \otimes \mathbb{R}$. A *polytope* $\Delta \subset N_\mathbb{R}$ is a convex hull of finite number of points in $N_\mathbb{R}$. A polytope is called *integral* iff these points lie in $N \otimes 1$. A polytope is called *primitive* if it is integral and its vertices are primitive. A Laurent polynomial is called *primitive* if its Newton polytope is primitive.

Definition 3.2. *The Minkowski sum* $\Delta_1 + \ldots + \Delta_k$ of polytopes $\Delta_1, \ldots, \Delta_k$ is the polytope $\{v_1 + \ldots + v_k | v_i \in \Delta_i\}$. An integral polytope is called *irreducible* if it can't be presented as a Minkowski sum of two non-trivial integral polytopes.

Remark 3.3. A Minkowski sum of integral polytopes is integral.

Definition 3.4. Consider an integral polytope $\Delta \subset \mathbb{Z}^n$. *A Minkowski presentation* of Δ is a presentation of each of its faces as a Minkowski sum of irreducible integral polytopes such that if a face Δ' lies in a face Δ then the intersections of Minkowski summands for Δ with Δ' give a presentation for Δ'.

Consider a Laurent polynomial $f \in \mathbb{C}[\mathbb{Z}^n]$. For any face Δ of Δ_f denote the sum of all monomials of f lying in Δ by f_Δ. The polynomial f is called a *Minkowski polynomial* if there exists a Minkowski presentation such that, for any face Δ of Δ_f with given Minkowski sum expansion $\Delta = \Delta_1 + \ldots + \Delta_k$, there are Laurent polynomials $f_{\Delta_i} \in \mathbb{C}[\mathbb{Z}^n]$ such that the coefficients of f_{Δ_i} at vertices of Δ_i are 1's and $f_\Delta = f_{\Delta_1} \cdot \ldots \cdot f_{\Delta_k}$.

Remark 3.5. Let e be an edge of a Minkowski Laurent polynomial of integral length n. Its unique Minkowski expansion to irreducible summands is the expansion to n segments of integral length 1. Thus the coefficient of the monomial associated to the i'th integral point of e (from any end) is $\binom{n}{i}$.

Remark 3.6. Toric Landau–Ginzburg models of Hori–Vafa type or toric Landau–Ginzburg models from [5] are Minkowski Laurent polynomials.

Example 3.7 (Ilten–Vollmert construction, [8]). Consider an integral polytope $\Delta \subset N = \mathbb{Z}^n$. Let the origin of N lie strictly inside Δ. Let $X = T_\Delta$ be the toric variety whose fan is the face fan for Δ. Denote the dual lattice to N by $M = N^\vee$. Put $N' = N \oplus \mathbb{Z}$, $M' = M \oplus \mathbb{Z}$. Let C be the cone generated by $(\Delta, 1)$. Then $X = Proj\,\mathbb{C}[C^\vee \cap M']$ with grading given by $d = (0, 1) \in M'$. For any primitive $r \in M'$, consider the map $r \colon N' \to \mathbb{Z}$. Let $L_r = ker(r)$. Let s_r be a retract (cosection) of the inclusion $i \colon L_r \to N'$, that is, a map $N' \to L_r$ such that $s_r i = Id_{L_r}$. It is unique up to translations along L_r. Let $C^+ = s_r(\{p \in C | \langle p, r \rangle = 1\})$, and $C^- = s_r(\{p \in C | \langle p, r \rangle = -1\})$ be two "slices" of C cut out by evaluating function at r.

Choose r such that $r = (r_0, 0) \in M'$ and such that C^- is a cone with its single vertex a lattice point. Consider a Minkowski decomposition $C^+ = C_1 + C_2$ to (possibly rational) polytopes such that for any vertex v of C^+, at least one of the corresponding vertices in C_1 and C_2 is a lattice point. Let D be the cone in $L_r \oplus \mathbb{Z}$ generated by $(C^-, 0)$, $(C_1, 1)$, and $(C_2, -1)$. Denote $X' = Proj\,\mathbb{C}[D^\vee \cap (L_r \oplus \mathbb{Z})^\vee)]$ where the grading is now given by $(s_r(d), 0)$.

Proposition 3.8 (Remark 1.8 and Theorem 4.4 in [8]). *There is an embedded degeneration of X' to X.*

Example 3.9 (Ilten). Let $\Delta \subset \mathbb{Z}^2$ be the convex hull of the points $(-1, 2)$, $(1, 2)$, and $(0, -1)$. Then $X = \mathbb{P}(1, 1, 4)$. Let $r = (0, 1, 0)$. Then s_r is given by the matrix

$$\begin{pmatrix} 1 & 0 & 0 \\ 0 & 1 & 1 \end{pmatrix}.$$

We are in the setup of Example 3.7 (see Figure 1). The vertex of $\{p \in C | \langle p, r \rangle = -1\}$ is $(0, -1, 1)$ and goes to a vertex $(0, 0)$ under s_r and the vertices of $\{p \in C | \langle p, r \rangle = 1\}$

are $(\pm\frac{1}{2}, 1, \frac{1}{2})$ and goes to vertices $(\pm\frac{1}{2}, \frac{3}{2})$ under s_r. That is, we have a Minkowski decomposition drawn on Figure 2.

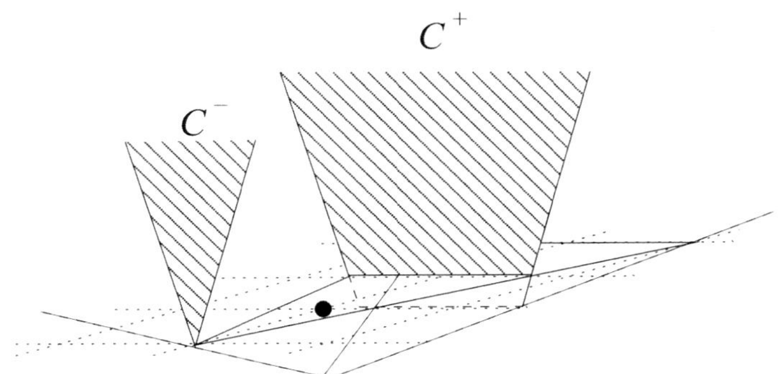

FIGURE 1. Deformation of $\mathbb{P}(1, 1, 4)$.

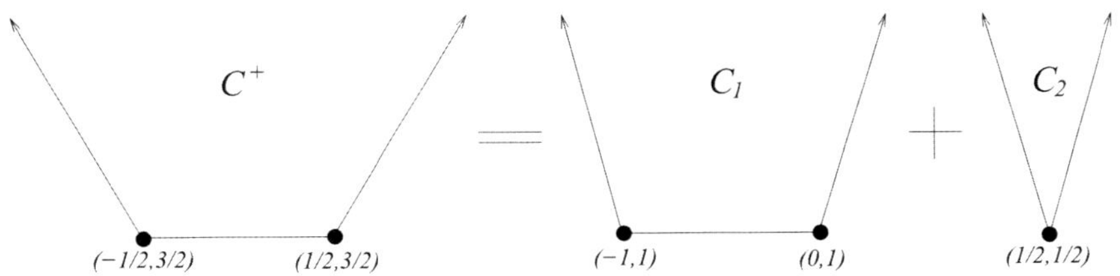

FIGURE 2. Decomposition of C^+.

The polytope for X' is a convex hull of points $(-1, 1)$, $(0, 1)$, and $(1, -2)$ since the second coordinate becomes to be equal to 1 not on $(C_2, -1)$ but on $(2C_2, -2)$. Its face fan is a fan of $\mathbb{P}^2$. Thus we get a deformation of $\mathbb{P}^2$ to $\mathbb{P}(1, 1, 4)$.

The following proposition shows that the degenerations given by Example 3.7 give cluster transformations for Minkowski polynomials.

Proposition 3.10. *Let $\Delta = \Delta_f$ be the Newton polytope of a Minkowski polynomial f. Let Δ' be a polytope obtained from Δ by the procedure described in Example 3.7 given by integral Minkowski summands agreeing with the Minkowski decompositions of the faces of Δ. Then $\Delta' = \Delta_{f'}$ for some Minkowski polynomial f'.*

Proof. Let $f \in \mathbb{C}[x_0^{\pm 1}, \ldots, x_n^{\pm 1}]$. After toric changes of variables we can assume that s_r is the projection on coordinates $x_1, \ldots, x_n$. Then

$$f = f_+(x_1, \ldots, x_n)x_0 + f_0(x_1, \ldots, x_n) + \frac{f_-(x_1, \ldots, x_n)}{x_0}.$$

As f is a Minkowski polynomial we have $f_+ = f_1 f_2$. Thus after change of variables $x_0 \to x_0/f_2$ we get a Minkowski polynomial

$$f' = f_1(x_1, \ldots, x_n)x_0 + f_0(x_1, \ldots, x_n) + \frac{f_-(x_1, \ldots, x_n)f_2(x_1, \ldots, x_n)}{x_0}$$

with Newton polytope Δ'. $\qquad\square$

Remark 3.11. Example 3.9 shows that the statement of Proposition 3.10 holds for non-integral case as well. This example is the first non-trivial cluster transformation given by Proposition 2.16.

Example 3.12. Let Δ be the convex hull of points $(-1, 1)$, $(1, 1)$, and $(0, -1)$. Then X is a quadratic cone $\mathbb{P}(1, 1, 2)$. (A unique) Minkowski polynomial for Δ is

$$f = \frac{(x+1)^2 y}{x} + \frac{1}{y}.$$

After cluster change of variables $y \to \frac{y}{x+1}$ we get a polynomial

$$\frac{(x+1)y}{x} + \frac{x+1}{y}.$$

It is (a unique) Minkowski polynomial for the polytope Δ' — the convex hull of points $(-1, -1)$, $(0, -1)$, $(1, -1)$, and $(0, -1)$. These points generate the fan of a smooth quadric X'.

4. Degenerations and wall crossings

In the previous section we have established certain combinatorial structures — cluster transformations connected to wall crossings. We will relate these combinatorial structures to the moduli space of stability conditions. We do this in two steps:

Step 1. First we relate the combinatorial structures to the "moduli space of Landau–Ginzburg models".

Step 2. Next we describe hypothetically how the "moduli space of Landau–Ginzburg models" fits in a "twistor family" with generic fiber the moduli space of stability conditions of a Fukaya–Seidel category.

We start with step one — collecting all Landau–Ginzburg models in a moduli space. The idea is to record wall crossings as relations in the mapping class group and then relations between relations and so on. This suggests a connection with Hodge theory and higher category theory. We will start with a rather simple approach which we will enhance later in order to serve our purposes. Nearly ten years ago it was discovered that, while the symplectic mapping class group of a curve equals the ordinary (oriented) mapping class group, these two groups differ greatly for higher dimensional symplectic manifolds. Understanding the structure of these groups has been a goal of many researchers in symplectic geometry. The initial purpose of the construction below was to obtain a presentation of the symplectic mapping class group of toric hypersurfaces. Along the way we have obtained a characterization of the zero fiber of a Stability Hodge Structure.

To explain our approach, we recall some notation and constructions. Assume $A \subset \mathbb{Z}^d$ is a finite set, X_A is the polarized toric variety associated to A with ample line bundle $\mathcal{L}$. In [9], the secondary polytope $\mathrm{Sec}(A)$ parameterizing regular subdivisions was constructed and shown to be the Newton polytope of the E_A determinant (a type of discriminant). We realize the toric variety associated to $\mathrm{Sec}(A)$ as the coarse moduli space of a stack $\mathcal{X}_{\mathrm{Sec}(A)}$ defined in [10]. We observe that the stack $\mathcal{X}_{\mathrm{Laf}(A)}$ constructed in [10] has a proper map π to $\mathcal{X}_{\mathrm{Sec}(A)}$ whose fibers are degenerations of X_A, and we constructed a polytope $\mathrm{Laf}(A)$ which is dual to the fan defining $\mathcal{X}_{\mathrm{Laf}(A)}$. The zero set $\mathcal{H}_{\mathrm{Sec}(A)}$ of a section of the associated line bundle parameterizes sections of $\mathcal{L}$ and degenerated sections are hypersurfaces in the associated degenerated toric variety. Upon restriction, we obtain a proper map $\pi : \mathcal{H}_{\mathrm{Sec}(A)} \to \mathcal{X}_{\mathrm{Sec}(A)}$ with non-singular fibers symplectomorphic to any non-degenerate section of $\mathcal{L}$.

Since $\pi : \mathcal{H}_{\mathrm{Sec}(A)} \to \mathcal{X}_{\mathrm{Sec}(A)}$ is a proper map, we may consider symplectic parallel transport of the non-singular fibers along paths in the complement of the zero set Z_A of the E_A determinant. Denote by H_p the fiber of π. We observe that the subset of the fibers meeting the toric boundary H are horizontal in the sense that if $q \in \partial H_p$ then the symplectic orthogonal $(T_q H_p)^{\perp \omega} \subset T_q(\partial \mathcal{H}_{\mathrm{Sec}(A)})$, where ω corresponds to a restriction of Fubini–Study metric. That is, parallel transport is a symplectomorphism that preserves the boundary of the hypersurfaces. Choosing a base point p of $\mathcal{X}_{\mathrm{Sec}(A)} \setminus Z_A$, we obtain a map from the based loop space $\rho : \Omega(\mathcal{X}_{\mathrm{Sec}(A)} \setminus Z_A) \to \mathrm{Symp}^{\partial}(H_p)$ and a group homomorphism

$$\rho_* : \pi_1(\mathcal{X}_{\mathrm{Sec}(A)} \setminus Z_A) \to \pi_0(\mathrm{Symp}^{\partial}(H_p)),$$

where $\pi_0(\mathrm{Symp}^{\partial}(H_p))$ is a mapping class group. However, from a field theory perspective, this homomorphism in imprecise; one should consider not only symplectomorphisms preserving the boundary, but also those that preserve the normal bundle of the boundary. In this way, we can glue two hypersurfaces together without creating an ambiguity in the symplectomorphism groups. We call such a symplectomorphism *boundary framed morphism* and denote the corresponding group $\mathrm{Symp}^{\partial,\mathrm{fr}}(H_p)$. For toric hypersurfaces, this group is a central extension of $\mathrm{Symp}^{\partial}(H_p)$. It is not generally the case, however, that parallel transport preserves the framing, but the change in framing can be controlled by keeping track of the homotopies in $\Omega(\mathcal{X}_{\mathrm{Sec}(A)} \setminus Z_A)$ or by passing to the loop space of an auxiliary real torus bundle $\mathcal{E} \to \mathcal{X}_{\mathrm{Sec}(A)} \setminus Z_A$, giving a homomorphism

$$\tilde{\rho}_* : \pi_1(\mathcal{E}) \to \pi_0(\mathrm{Symp}^{\partial,\mathrm{fr}}(H_p)).$$

In many cases, this homomorphism is surjective.

The stack $\mathcal{X}_{\mathrm{Sec}(A)}$ is as complicated combinatorially as the secondary polytope $\mathrm{Sec}(A)$, which is computationally expensive to describe. While the Newton polytope of E_A was found in [9], Z_A is far from smooth and there are open questions about its singular structure. We bypass these difficulties by considering only the lowest dimensional boundary strata of $\mathcal{X}_{\mathrm{Sec}(A)}$ where non-trivial behavior occurs. Thus the first and main case we examine are the one dimensional boundary strata of $\mathcal{X}_{\mathrm{Sec}(A)}$. Combinatorially, these are known as circuits.

A *circuit* A is a collection of $d+2$ points in $\mathbb{Z}^d$, such that there are exactly two coherent triangulations (see [9]) of A, so the secondary polytope is a line segment and the secondary stack a weighted projective line $\mathbb{P}(a,b)$. Z_A is either two or three points; two of the points are the equivariant orbifold points $\{0, \infty\}$ and the possible third is an interior point. Both the constants, a, b and the number of points in Z_A depends on the convex hull and affine positioning of A — for more details see [2]. When Z_A consists of three points, their complement retracts onto a figure eight and the fundamental group is free on two letters. In this case, we have the based loops $\delta_1, \delta_2, \delta_3 = \delta_2^{-1}\delta_1^{-1}$ encircling the three points. The symplectic monodromy $T_i = \tilde{\rho}_*(\delta_i)$ is computable from known results in symplectic geometry as either spherical Dehn twists or as twists about a tropical decomposition. The image via $\tilde{\rho}_*$ gives the relation

$$T_1 T_2 T_3 = T_{\partial H_p}, \tag{1}$$

where $T_{\partial H_p}$ is the central element determined by twisting the framing about the toric boundary. One of the most elementary examples is $X_A = \mathbb{P}^1 \times \mathbb{P}^1$ with polarization $\mathcal{O}(1,1)$, and the circuit is the four vertices of a unit square with the two diagonal triangulations. Here the hypersurface is $\mathbb{P}^1$ with four boundary points and the relation obtained above yields a classical relation in the mapping class group called the *Lantern relation.*

When Z_A consists of two points, one is an orbifold point and the other is a point with trivial stabilizer. If δ_1, δ_2 are based paths encircling Z_A and T_1, T_2 are the associated symplectomorphisms, we obtain a relation

$$(T_1 T_2)^a = T_{\partial H_p}. \tag{2}$$

A basic example of this relation arises as the homological mirror to $\mathbb{P}^2$ which is the set $A = \{(0,0), (1,0), (0,1), (-1,-1)\}$. The constant a occurring above is 3 and the relation is in fact another classical mapping class group relation known as the *star relation.*

We call the boundary framed, symplectic mapping class group relation occurring in equations 1 and 2 *the circuit relation.* In general, any complex line in $\mathcal{X}_{\mathrm{Sec}(A)}$ yields a relation in $\mathrm{Symp}^{\partial,\mathrm{fr}}(H_p)$ by homotoping the product of all the loops around the intersections with Z_A to the identity. However, each such line can be degenerated to a chain of equivariant lines which are precisely circuits supported on A. Thus every relation obtained this way can be thought of as arising from a composition of circuit relations. As we saw in the previous two sections Landau–Ginzburg mirrors of Fano manifolds are fibrations of Calabi–Yau hypersurfaces. Therefore the above simple examples generalize to

Theorem 4.1 ([2]). *Landau–Ginzburg mirrors of Fano manifolds can be obtained by a superposition of circuits described above.*

Interpreting Landau–Ginzburg models as lines in the secondary stack we get

Theorem 4.2 ([2]). $\mathcal{X}_{Sec(A)}$ *can be seen as moduli space of Landau–Ginzburg models. In particular some wall crossings correspond to passing through Z_A.*

These two theorems complete Step 1.

5. Wall crossings and Stability Hodge Structures

We move to Step 2, building a "twistor family" with generic fiber the moduli space of stability conditions for Fukaya–Seidel categories — see [11].

Stability Hodge Structures. We start with Stability Hodge Structures, an artifact of Donaldson–Thomas (DT) invariants. We will mainly consider Fukaya–Seidel categories but discussion in this section applies in general.

The theory of Donaldson–Thomas invariants and wall crossing has become a central subject of Geometry and Physics. In a nutshell DT invariants are virtual numbers of stable objects in three dimensional Calabi–Yau category. Kontsevich and Soibelman suggested Donaldson–Thomas invariants applicable to triangulated category and Bridgeland stability conditions — a refined version of so called *motivic Donaldson–Thomas invariants* — MDT. The wall crossing formulae (WCF) of MDT are expressed in terms of factorization of quantum torus. A connection with nonabelian Hodge structures comes naturally here. WCF for the Hitchin system is connected to ODE with small parameter and its asymptotic behavior. In fact the WCF relates to Stokes data at infinity for this ODE and connects with the work of Ecalle and Voros on resurgence.

We will introduce a new geometric structure which seems to be present in many of above considerations — Stability Hodge Structures. These structures seem to have a huge potential of geometric applications some of which we discuss.

The moduli space of stability conditions of a category C is very complicated with possibly fractal boundary. In the case of derived category of Calabi–Yau manifolds of dimension three and higher there is not any hypothetical description. Still HMS predicts that the moduli space of mirror dual Calabi–Yau manifolds is embedded in a locally closed cone in the moduli space of stability conditions of a category C. So it is a big open question how to characterize Hodge structures corresponding to mirror duals. Classically the moduli space of pure Hodge structures has a compactification by Mixed Hodge Structures (MHS). So it is natural to study limiting Donaldson–Thomas invariants and relate to WCF.

In the case of three-dimensional Calabi–Yau manifolds there are different types of MHS. The cusp case — the deepest degeneration — corresponds to t-structures which is an extension of Tate motives. As a result we take a generating series of Donaldson–Thomas rank one torsion free invariants. It is known that in this case this generating series (modulo change of coordinates) is the classical Gromov–Witten series which satisfies the holomorphic anomaly equation. This translates into automorphic property for the DT generating function. We expect that automorphic property holds for higher ranks and plan to study it and show that WCF is necessary to assemble limiting data.

A different MHS corresponds to conifold points and non-maximal degeneration points. The wall crossings and DT data give a family of Integrable Systems in the following way. The vanishing cycles Γ_{short} and the monodromy define a quotient category $\mathcal{T}/\mathcal{A}$ with the following sequence on the level of K–theory:

$$\Gamma_{short} \to K_0(\mathcal{T}) \to K_0(\mathcal{T}/\mathcal{A}).$$

Using the Kontsevich–Soibelman noncommutative torus approach we define a super-scheme

$$\mathbb{G} = \oplus_{p \in \Gamma_{short}} \mathbb{G}_p \to T_{non}.$$

Consider the zero grade $\mathbb{G}_0$ of $\mathbb{G}$ over $\mathbb{Z}$. The global sections of $\mathbb{G}_0$ define the Betti moduli space — an integrable system

$$\Gamma(G_0) = \oplus \mathcal{O}(M_j).$$

In order to consider the interaction with the rest of the category we include global WCF. In this case we obtain a torus action, which produces a stack over the Betti moduli space:

$$X/(\mathbb{C}^*)^{\times n} \to M_1 \times M_2 \times \ldots \times M_k.$$

All these stacks fit in a constructible sheaf.

To summarize we give a provisional definition, which covers the cases of Bridgeland, geometric (volume forms), and generalized (log forms) stability conditions:

Definition 5.1. A *Stability Hodge Structure (SHS)* for a Fukaya–Seidel category $\mathcal{F}$ is the following data:

   i) The moduli space of stability conditions S for $\mathcal{F}$.

  ii) Divisor D at infinity giving a partial compactification of S and parametrizing the degenerated limiting stability conditions — stability conditions for quotient categories, the category factored by the objects (vanishing cycles) on which stability conditions vanish.

 iii) Besides the degeneration we record the WCF — all recorded together. Over each point of D we put the Betti moduli space locally produced by WCF. All these moduli space fit in a constructible sheaf over S.

Let us illustrate these structures through two examples. We start with the category $\widetilde{\mathbb{A}}_2$ — the Fukaya category of the conic bundle $\{uv = y^2 - x^3 - ax - b\}$, $a, b \in \mathbb{C}$. In this case, the Stability Hodge Structure is a sheaf over $\mathbb{C}^2$ with coordinates a, b.

The points of the discriminant parameterize limiting stability conditions. The fibers are Betti moduli spaces of vanishing cycles which generically over the discriminant are the affine surface $z(1 - xy) = 1$. The special fiber over the cusp is the moduli space $M_{0,5}$ of rank two bundles over the projective line with one irregular singularity and five Stokes directions at infinity (see Figure 3).

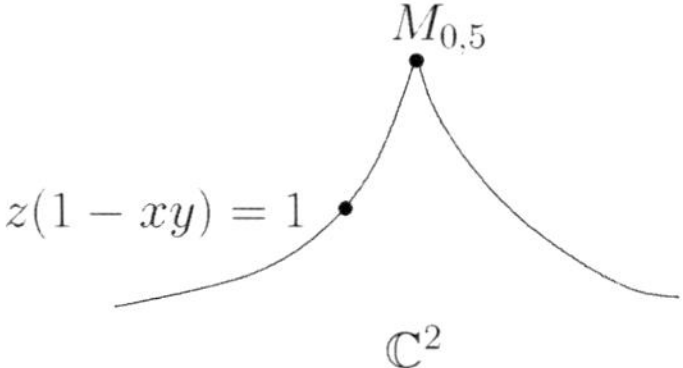

FIGURE 3. Compactification of moduli space of stability conditions for a category $\widetilde{\mathbb{A}}_2$.

A different example is the Fukaya–Seidel category $\mathbb{A}_4$. We start with a generic polynomial $p \in \mathbb{C}[z]$ of degree 5. It defines a Riemann surface $C = \{p(z) = w\}$ and 5:1-covering $\varphi \colon C \to \mathbb{C}$. The ramification locus for φ are 4 points $p_1, \ldots, p_4$— roots of p'. Consider 4 paths $l_1, \ldots, l_4$ from p_i's to infinity. The polynomial p is generic, so the ramification is as simple as it can be and $\varphi^{-1}(l_i)$ are thimbles covering l_i's 2:1. They generate a Fukaya category for C and correspond to vertices of the $\mathbb{A}_4$ quiver. "Neighbor" thimbles intersect at infinity: i-th one intersects $(i+1)$-th at one point. These intersections correspond to arrows between vertices in the quiver.

In this example the divisor D at infinity parameterizes the semiorthogonal decompositions of the $\mathbb{A}_4$ category. The fibers of the constructible sheaf are moduli spaces of stability conditions for $\mathbb{A}_3 \times \mathbb{A}_1$ categories. Similarly on the singular points of D we get as fibers moduli spaces of stability conditions for $\mathbb{A}_2 \times \mathbb{A}_2$ categories. This leads to a rich mixed Hodge theory structure associated with D and monodromy action around it. In the next section we will see that in the limit the stability conditions behave as coverings so the above picture fits. This monodromy relates to the wall-crossings changes. In particular it sends the preferred set of thimbles generating the $\mathbb{A}_4$ category from a generator consisting of the sum of 4 thimbles $G = L_1 + L_2 + L_3 + L_4$ (with $Hom(L_i, L_{i+1})$ of rank 1) to $G' = L' + L_1 + L_3 + L_4$ by a mutation. This mutation reduces the generation time (see Section 7) from $t(G) = 3$ to $t(G') = 2$. We will represent it as an invariant of of Stability Hodge Structures in Section 7.

In the next section we build a twistor type of family where the generic fiber is a SHS.

6. Higgs bundles and stability conditions — analogy

In this section we proceed describing the analogy between Nonabelian and Stability Hodge Structures. We build the "twistor" family so that the fiber over zero is the "moduli space" of Landau–Ginzburg models and the generic fiber is the Stability Hodge Structure defined above.

Noncommutative Hodge theory endows the cohomology groups of a dg-category with additional linear data — the noncommutative Hodge structure — which records important information about the geometry of the category. However, due to their linear nature, noncommutative Hodge structures are not sophisticated enough to codify the full geometric information hidden in a dg-category. In view of the homological complexity of such categories it is clear that only a subtler non-linear Hodge theoretic entity can adequately capture the salient features of such categorical or noncommutative geometries. In this section by analogy with "classical nonabelian Hodge theory" we construct and study from such a perspective a new type of entity of exactly such type — the Stability Hodge Structure associated with a dg-category.

As the name suggests, the SHS of a category is related to the Bridgeland stabilities on this category. The moduli space Stab_C of stability conditions of a triangulated dg-category C is, in general, a complicated curved space, possibly with fractal boundary. In the special case when C is the Fukaya category of a Calabi–Yau threefold, the space Stab_C admits a natural one-parameter specialization to a much simpler space S_0. Indeed, HMS predicts that the moduli space of complex structures on the Calabi–Yau threefold maps to a Lagrangian subvariety $\mathsf{Stab}_C^{\mathrm{geom}} \subset \mathsf{Stab}_C$. (Recall the holomorphic volume form and integrating it defines a stability condition and its charges.) The idea is now to linearize Stab_C along $\mathsf{Stab}_C^{\mathrm{geom}}$, i.e., to replace Stab_C with a certain discrete quotient S_0 of the total space of the normal bundle of $\mathsf{Stab}_C^{\mathrm{geom}}$ in Stab_C. Specifically, by scaling the differentials and higher products in C, one obtains a one parameter family of categories C_λ with $\lambda \in \mathbb{C}^*$, and an associated family $\mathsf{S}_\lambda := \mathsf{Stab}_{C_\lambda}$, $\lambda \in \mathbb{C}^*$ of moduli of stabilities. Using holomorphic sections with prescribed asymptotic at zero one can complete the family $\{\mathsf{S}_\lambda\}_{\lambda \in \mathbb{C}^*}$ to a family $\mathsf{S} \to \mathbb{C}$ which in a neighborhood of $\mathsf{Stab}_C^{\mathrm{geom}}$ behaves like a standard deformation to the normal cone. The space S_0 is the fiber at 0 of this completed family and conjecturally $\mathsf{S} \to \mathbb{C}$ is one chart of a twistor-like family $\mathcal{S} \to \mathbb{P}^1$ which is by definition *the Stability Hodge Structure associated with C.*

Stability Hodge Structures are expected to exist for more general dg-categories, in particular for Fukaya–Seidel categories associated with a superpotential on a Calabi–Yau space or with categories of representations of quivers. Moreover, for special non-compact Calabi–Yau 3-folds, the zero fiber S_0 of a Stability Hodge Structure can be identified with the Dolbeault realization of a nonabelian Hodge structure of an algebraic curve. This is an unexpected and direct connection with Simpson's nonabelian Hodge theory which we exploit further suggesting some geometric applications.

We briefly recall nonabelian Hodge theory settings. According to Simpson we have one parametric twistor family such that the fiber over zero is the moduli space of Higgs bundles and the generic fiber is the moduli space of representations of the fundamental group — M_{Betti}.

In this section we state that we expect similar behavior of moduli space of stability conditions. The moduli space of stability conditions of a Fukaya–Seidel category can be included in a one parameter twistor family, and we describe the fiber over zero in details in the next subsection.

We give an example:

Example 6.1 ("twistor" family for Stability Hodge Structures for the category $\mathbb{A}_n$). We will give a brief explanation the calculation of the "twistor" family for the SHS for the category $\mathbb{A}_n$. We start with the moduli space of stability conditions for the category $\mathbb{A}_n$, which can be identified with differentials $e^p dz$, where $p \in \mathbb{C}[z]$ is a generic polynomial of degree $n + 1$, see [1].

Let us denote one particular holomorphic form $e^p dz$ by Vol. Locally there exists a holomorphic coordinate w such that $Vol = dw$. Geodesics in the metric $|Vol|^2$ are the straight real lines in the coordinate w, the same as real lines on which Vol has constant phase. Therefore they are special Lagrangians for Vol (and in fact for any real symplectic structure).

Observe that these geodesics are asymptotic to infinity because the integral of $|Vol| = e^{Re(p)}|dz|$ absolutely converges on them hence $Re(p)$ approaches infinity, as these lines are noncompact in the uncompactified plane z, and therefore $|z|$ goes to infinity. To compensate infinite length in the usual metric $|dz|^2$ we use the fact that $e^{Re(p)}$ converges to zero iff $Re(p)$ converges to minus infinity.

So after completion in the metric defined above (so the vertices are in the finite part now) we enhance the polygon by assigning angles and lengths. These enhanced polygons record our stability conditions. Indeed we have $(2(n+1)-3)$-dimensional space of polygons plus one global angle — it is a real $2n$ dimensional space. In Example 6.2 we give a simple example illustrating the polygons for the category $\mathbb{A}_2$ and a wall crossing phenomenon. The stable objects correspond to edges and diagonals. In the picture in the example we lose one stable object while crossing a wall.

Example 6.2 (stability for $\mathbb{A}_2$). For $\mathbb{A}_2$ category we have $\deg p = 3$. The left part of Figure 4 represents two of the stable objects for the $\mathbb{A}_2$ category. The third stable object is the third edge of the triangle. The wall crossing makes the angle between the first two edges bigger then π and as a result the third edge is not a stable object any more.

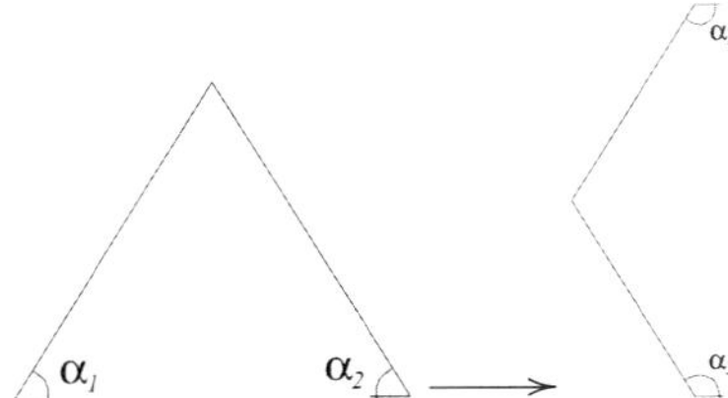

FIGURE 4. Stability conditions for the $\mathbb{A}_2$ category.

Now we consider the "twistor" family — the limit of $e^{p(z)/u}dz$, where u is a complex number tending to $\mathbb{C}$. Geometrically limit differential can be identified with graphs — see Example 6.3.

Example 6.3 (limit of $e^{p/u}dz$). Take a limit of $e^{p/u}dz$ with u tending to zero. The limits of polygons are graphs. We record the length, angle, and monodromy and this defines a covering of the complex plane. Thus this construction identifies a limit of moduli space of stability conditions for $\mathbb{A}_n$ category with some Hurwitz subspace — a subscheme of coverings. In particular, these two spaces have the same number of components. Figure 5 represents a procedure of associating the monodromy of the covering to the vertices of the graph.

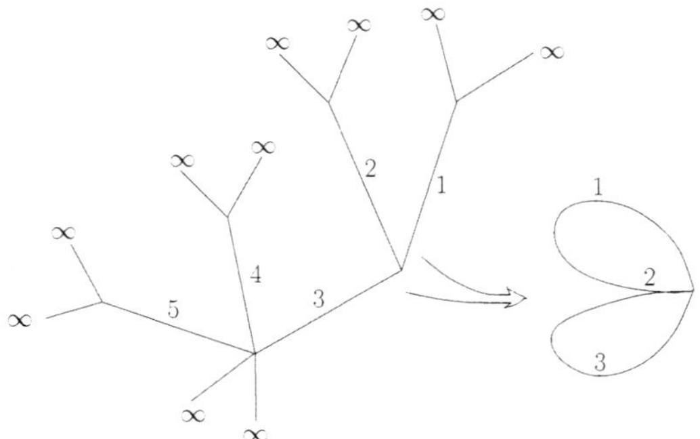

FIGURE 5. Building coverings out of limit.

Remark 6.4. Similarly one can compute the "twistor" family for the equivariant $\mathbb{A}_n$ category and see appearance of gaps in spectra in connection with the weight filtration of completions of special local rings — see Section 7. Observe that the idea of coverings brings the Fukaya category of a Riemann surface of genus g very close to the $\mathbb{A}_{2g+1}$ category. Also product of Fukaya categories of curves in combination with Luttinger surgeries gives many opportunities for stability conditions with many components as well as many possibilities for the behavior of gaps and spectra. The interplay between coverings and stability conditions suggests that one can have symplectic manifolds with the same Fukaya categories but different moduli of stability conditions. We conjecture that the moduli spaces of coverings obtained near different cusps being different algebraically should imply that this different manifolds are nonsymplectomorphic.

The fiber over zero. The fiber over zero (described in what follows) plays an analogous role to the moduli space of Higgs bundles in Simpson's twistor family in the theory of nonabelian Hodge structures. Constructing it amounts to a repetition of our construction in Section 4 from a new perspective and enhanced with more structure.

The $\mathbb{A}_n$ example considered above is a simple example of more general Fukaya–Seidel categories that arise in Homological Mirror Symmetry. Stability conditions associated to the Fukaya–Seidel category are closely related to the complex deformation parameters, i.e., the moduli space of Landau–Ginzburg models. We begin by recalling the general setup in the case of Landau–Ginzburg models. The prescription given by Batyrev, Borisov, Hori, Vafa in [12], [13] to obtain homological mirrors for toric Fano varieties is perfectly explicit and provides a reasonably large set of examples to examine. We recall that if Σ is a fan in $\mathbb{R}^n$ for a toric Fano variety X_Σ, then the homological mirror to the B model of X_Σ is a Landau–Ginzburg model $w : (\mathbb{C}^*)^n \to \mathbb{C}$ where the Newton polytope Q of w is the convex hull of generators of rays of Σ. In fact, we may consider the domain $(\mathbb{C}^*)^n$ to occur as the dense orbit of a toric variety X_A, where A is $Q \cap \mathbb{Z}^n$ and X_A indicates the polytope toric construction. In this setting, the function w occurs as a pencil $V_w \subset H^0(X_A, L_A)$ with fiber at infinity equal to the toric boundary of X_A. Similar construction works for generic non-toric Fanos. In this paper we work with the directed Fukaya category associated to the superpotential w — Fukaya–Seidel categories. To build on the discussion above, we discuss here Fukaya–Seidel categories in the context of stability conditions. The fiber over zero corresponds to the moduli of complex structures. If X_A is toric, the space of complex structures on it is trivial, so the complex moduli appearing here are a result of the choice of fiber $H \subset X_A$ and the choice of pencil w respectively. The appropriate stack parameterizing the choice of fiber contains the quotient $[U/(\mathbb{C}^*)^n]$ as an open dense subset where U is the open subset of $H^0(X_A, L_A)$ consisting of those sections whose hypersurfaces are nondegenerate (i.e., smooth and transversely intersecting the toric boundary) and $(\mathbb{C}^*)^n$ acts by its action on X_A. To produce a reasonably well-behaved compactification of this stack, we borrow from the works of Alexeev ([14]), Gelfand, Kapranov, and Zelevinsky ([9]), and Lafforgue ([10]) to construct the stack $\mathcal{X}_{\mathrm{Sec}(A)}$ with universal hypersurface stack $\mathcal{X}_{Laf(A)}$. We quote the following theorem which describes much of the qualitative behavior of these stacks:

Theorem 6.5 ([2]).   i) *The stack $\mathcal{X}_{Sec(A)}$ is a toric stack with moment polytope equal to the secondary polytope $Sec(A)$ of A.*

   ii) *The stack $\mathcal{X}_{Laf(A)}$ is a toric stack with moment polytope equal to the Minkowski sum $Sec(A) + \Delta_A$ where Δ_A is the standard simplex in $\mathbb{R}^A$.*

   iii) *Given any toric degeneration $F : Y \to \mathbb{C}$ of the pair (X_A, H), there exists a unique map $f : \mathbb{C} \to \mathcal{X}_{Sec(A)}$ such that F is the pullback of $\mathcal{X}_{Laf(A)}$.*

We note that in the theorem above, the stacks $\mathcal{X}_{Laf(A)}$ and $\mathcal{X}_{Sec(A)}$ carry additional equivariant line bundles that have not been examined extensively in existing literature, but are of great geometric significance. The stack $\mathcal{X}_{Sec(A)}$ is a moduli stack for toric degenerations of toric hypersurfaces $H \subset X_A$. There is a hypersurface $\mathcal{E}_A \subset \mathcal{X}_{Sec(A)}$ which parameterizes all degenerate hypersurfaces. For the Fukaya category of hypersurfaces in X_A, the complement $\mathcal{X}_{Sec(A)} \setminus \mathcal{E}_A$ plays the role of the classical stability conditions, while including $\mathcal{E}_A$ incorporates the compactified version where MHS come into effect. We predict that the walls of the stability conditions occurring in this setup are seen as components of the tropical amoeba defined by the principal A-determinant E_A.

To find the stability conditions associated to the directed Fukaya category of (X_A, w), one needs to identify the complex deformation parameters associated to this model. In fact, these are precisely described as the coefficients of the superpotential, or in our setup, the pencil $V_w \subset H^0(X_A, w)$. Noticing that the toric boundary is also a toric degeneration of the hypersurface, we have that the pencil V_w is nothing other than a map from $\mathbb{P}^1$ to $\mathcal{X}_{\mathrm{Sec}(A)}$ with prescribed point at infinity. If we decorate $\mathbb{P}^1$ with markings at the critical values of w and ∞, then we can observe such a map as an element of $\mathcal{M}_{0,Vol(Q)+1}(\mathcal{X}_{\mathrm{Sec}(A)},[w])$ which evaluates to $\mathcal{E}_A$ at all points except one and ∂X_A at the remaining point. We define the cycle of all stable maps with such an evaluation to be $\mathcal{W}_A$ and regard it as the appropriate compactification of complex structures on Landau–Ginzburg A-models. Applying techniques from fiber polytopes we obtain the following description of $\mathcal{W}_A$:

Theorem 6.6 ([2]). *The stack $\mathcal{W}_A$ is a toric stack with moment polytope equal to the monotone path polytope of $Sec(A)$.*

The polytope occurring here is not as widely known as the secondary polytope, but occurs in a broad framework of so called iterated fiber polytopes introduced by Billera and Sturmfels.

In addition to the applications of these moduli spaces to stability conditions, we also obtain important information on the directed Fukaya categories and their mirrors from this approach. In particular, the above theorem may be applied to computationally find a finite set of special Landau–Ginzburg models $\{w_1,\ldots,w_s\}$ corresponding to the fixed points of $\mathcal{W}_A$ (or the vertices of the monotone path polytope of $Sec(A)$). Each such point is a stable map to $\mathcal{X}_{\mathrm{Sec}(A)}$ whose image in the moment space lies on the 1-skeleton of the secondary polytope. This gives a natural semiorthogonal decomposition of the directed Fukaya category into pieces corresponding to the components in the stable curve which is the domain of w_i. After ordering these components, we see that the image of any one of them is a multi-cover of the equivariant cycle corresponding to an edge of $Sec(A)$. These edges are known as circuits in combinatorics (see [2]).

Now we put this moduli space as a "zero fiber" of the "twistor" family of moduli family of stability conditions.

We do this in two steps:

1. The following theorem suggests the existence of a formal moduli space M of Landau–Ginzburg models $f\colon \overline{Y} \to \mathbb{C}\mathbb{P}^1$.

Theorem 6.7 (see [1]). *There exists a formal moduli space M determined by the solutions of the Maurer–Cartan equations for the following dg-complex:*

$$\cdots \longleftarrow \underset{-3}{\Lambda^3 T_{\overline{Y}}} \longleftarrow \underset{-2}{\Lambda^2 T_{\overline{Y}}} \longleftarrow \underset{-1}{T_{\overline{Y}}} \longleftarrow \underset{0}{\mathcal{O}_{\overline{Y}}} \longleftarrow 0$$

In the above complex the differential is df and we can restate it by saying that this complex determines deformations of the Landau–Ginzburg model, and these deformations are unobstructed. We also have a $\mathbb{C}^*$-action on M with fixed points corresponding to limiting stability conditions — see [1].

Over the moduli space M defined above we have a variation of Hodge structures defined by the cohomologies of the perverse sheaf of vanishing cycles over Y. This defines local system V over M and its compactification.

116

Conjecture 6.8 (see [1]). *The relative completion with respect of V in the fixed points of the $\mathbb{C}^*$-action on the compactification of M has a mixed Hodge structure.*

2. The above moduli space is too big. So we will cut its dimension down to the moduli space of stability conditions. We introduce a new moduli space which embeds in M.

We study deformations of $\overline{Y} \to \mathbb{CP}^1$ with "fixing the fiber at infinity". Deformation of a smooth variety $\overline{Y}$ with fixed $\mathbb{CP}^1$ is controlled by the following sheaf of dg Lie algebras on $\overline{Y}$:

$$T_{\overline{Y}} \to f^* T_{\mathbb{CP}^1}$$

(the differential is the tangent map).

By fixing the fiber at infinity we get a subsheaf of dg Lie algebras

$$T_{\overline{Y}, Y_\infty} \to f^* T_{\mathbb{CP}^1, \infty}.$$

Theorem 6.9 ([1]). *The subsheaf of dg Lie algebras*

$$T_{\overline{Y}, Y_\infty} \to f^* T_{\mathbb{CP}^1, \infty}$$

determines a smooth moduli stack. Its dimension is equal to the dimension of the moduli space of stability conditions.

A geometric realization of this moduli space, which embeds in M was described above. We will denote it by $M(\mathbb{P}^1, CY)$ (or $M(\mathbb{P}^k, CY)$ for multipotential Landau–Ginzburg models).

Remark 6.10. We can consider a bigger moduli space by fixing the vector fields only over a part of the divisor at infinity. This corresponds to taking a Landau–Ginzburg model through a point of non maximal degeneration. This defines a bigger moduli space of stability conditions with more stable objects.

Remark 6.11. The moduli spaces we discuss could have many components. Such a phenomenon would have many interesting implications. It produces possibilities of many new birational and symplectic invariants.

In the same way as the fixed point set under the $\mathbb{C}^*$-action plays an important role in describing the rational homotopy types of smooth projective varieties we study the fixed points of the $\mathbb{C}^*$-action on F and derive information about the homotopy type of a category. In the rest of the paper we will denote $\mathcal{W}_A$ by $M(\mathbb{P}^1, \mathcal{X}_{\mathrm{Sec}(A)})$ (or $M(\mathbb{P}^k, CY)$) in order to stress the connection with Landau–Ginzburg models (here CY denotes the moduli space of Calabi–Yau mirrors to the anticanonical section of the Fano manifold we consider).

7. Spectra and holomorphic convexity

In this section we explain briefly how Orlov spectra are related to Stability Hodge Structures.

Recall that noncommutative Hodge structures were introduced by Kontsevich and Katzarkov and Pantev [15] as means of bringing the techniques and tools of Hodge theory into the categorical and noncommutative realm. In the classical setting, much of the information about an isolated singularity is recorded by means of the Hodge spectrum, a set of rational eigenvalues of the monodromy operator. The Orlov spectrum (defined below), is a categorical analogue of this Hodge spectrum appearing in the work of Orlov and Rouquier. The missing numbers in the spectra are called gaps.

Let $\mathcal{T}$ be a triangulated category. For any $G \in \mathcal{T}$ denote by $\langle G \rangle_0$ the smallest full subcategory containing G which is closed under isomorphisms, shifting, and taking finite direct sums and summands. Now inductively define $\langle G \rangle_n$ as the full subcategory of objects, B, such that there is a distinguished triangle, $X \to B \to Y \to X[1]$, with $X \in \langle G \rangle_{n-1}$ and $Y \in \langle G \rangle_0$, and direct summands of such objects.

Definition 7.1. Let G be an object of a triangulated category $\mathcal{T}$. If there is an n with $\langle G \rangle_n = \mathcal{T}$, we set

$$t(G) := \min \left\{ n \geq 0 \mid \langle G \rangle_n = \mathcal{T} \right\}.$$

Otherwise, we set $t(G) := \infty$. We call $t(G)$ the *generation time* of G. If $t(G)$ is finite, we say that G is a *strong generator*. The *Orlov spectrum* of $\mathcal{T}$ is the union of all possible generation times for strong generators of $\mathcal{T}$. The *Rouquier dimension* is the smallest number in the Orlov spectrum. We say that a triangulated category $\mathcal{T}$, has a *gap* of length s, if a and $a + s + 1$ are in the Orlov spectrum but r is not in the Orlov spectrum for $a < r < a + s + 1$.

The first connection to Hodge theory appears in the form of the following theorem:

Theorem 7.2 ([16]). *Let X be an algebraic variety possessing an isolated hypersurface singularity. The Orlov spectrum of the category of singularities of X is bounded by twice the embedding dimension times the Tjurina number of the singularity.*

After this brief review of the theory of spectra and their gaps we connect them with SHS. Let $SHS(X)$ be the Stability Hodge Structure of $D^b(X)$ for a given Fano variety X, $M(\mathbb{P}^1, CY)$ be its zero fiber.

Conjecture 7.3 ([1]). *Let p be a point of the divisor D at infinity of the compactification of $M(\mathbb{P}^1, CY)$. The mixed Hodge structures on the completion of the local ring O_p, where p runs over all commponnents of D, determines the spectrum of $D^b(X)$.*

Remark 7.4. The above considerations suggests the existence of a Riemann–Hilbert correspondence for $SHS(X)$ for a Fano variety X as well as deep and interesting analytical interpretation of it by analogy with Yang–Mills–Higgs equations.

As a consequence of the above conjecture we have that SHS satisfy two important properties — functoriality and strictness. We arrive at:

Conjecture 7.5. *The infinite chain condition ([3]) can be ruled out for the universal coverings of smooth projective surfaces.*

This is the strongest obstruction to Shafarevich conjecture [3] and SHS gives an approach proving that universal coverings of smooth projective varieties are holomorphically convex.

Observe that the "twistor" family of compactified SHS depends on the choice of Landau–Ginzburg model and still computes some purely categorical invariants. It is natural to ask whether this family rigidifies the data. In particular we pose:

Question 7.6. Does the "twistor" family of compactified SHS of bounded derived category of coherent sheaves of a smooth projective variety X recover the fundamental group of X?

8. Multipotential Landau–Ginzburg models and Hodge structures.

In this section we extend the correspondence among categories and Stability Hodge Structures further. We underscore the idea that rich geometry of the Landau–Ginzburg models gives a possibility of constructing interesting Stability Hodge Structures with many filtrations.

8.1. Multipotential Landau–Ginzburg model for cubic fourfold

We describe fiberwise compactifications of multipotential Landau–Ginzburg models for the cubic fourfold X. This example is representative and illustrates what we mean by a multipotential Landau–Ginzburg model in general.

The Hori–Vafa toric Landau–Ginzburg for X is

$$w = \frac{(x+y+1)^3}{xyt_1t_2} + t_1 + t_2.$$

The cubic fourfold is of index 3. So there are two decompositions of its anticanonical divisor: $3H = H + H + H$ and $3H = 2H + H$. Multipotential Landau–Ginzburg models correspond to such decompositions.

First we describe compactification for the first decomposition. We have the family

$$\frac{(x+y+1)^3}{xyt_1t_2} = w_1, \qquad t_1 = w_2, \qquad t_2 = w_3,$$

where w_i's are complex parameters. After compactifying we get the family

$$(x+y+z)^3 = w_1w_2w_3xyz$$

of elliptic curves over $\mathbb{C}^3$. After blowing up the point $(0,0,0)$ we get a divisor over this point. After that we resolve the rest of the singularities. The restriction of our family to planes $w_j = const \neq 0$ is the Landau–Ginzburg model for cubic threefold so we get the following configuration of singularities.

(1) Ordinary double points along the surface $w_1w_2w_3 = 27$.
(2) 7 lines forming a diagram of type $\widetilde{\mathbb{E}}_6$ over planes $w_1 = 0$, $w_2 = 0$, and $w_3 = 0$.
(3) 5 surfaces over axes w_1, w_2, w_3.
(4) A divisor over (0,0,0).

After projection on the diagonal $\mathbb{C}^3 \to \mathbb{C}$ we get a fiberwise open part of the usual Landau–Ginzburg model for cubic fourfold. Its fiber over zero consists of the divisor described above and an elliptic fibration over the plane passing through the origin and orthogonal to the diagonal. The intersection of these divisors is an elliptic K3 surface with 3 fibers of type $\widetilde{\mathbb{E}}_6$ corresponding to intersections of this orthogonal plane with planes $w_1 = 0$, $w_2 = 0$, and $w_3 = 0$.

Now we describe multipotential Landau–Ginzburg model for the second case $2H + H$. We have the family

$$\frac{(x + y + z)^3}{xyzt_1t_2} = w_1, \qquad t_1 + t_2 = w_2.$$

In other words,

$$(x + y + z)^3 = w_1(w_2 - t)txyz$$

(we denote t_1 by t for simplicity).

This family of surfaces can be obtained from the decomposition $H + H + H$ by a projection along $w_2 + w_3 = 0$. Indeed, the equation of this family over $\mathbb{C}^2$ can be obtained from the equation for the family over $\mathbb{C}^3$ by the coordinate change $w_2 + w_3 \to w_2$, $w_3 \to t$.

So the singularities are the following.

(1) Ordinary double points along a curve.
(2) 5 surfaces over the axis $w_2 = 0$.
(3) 17 surfaces over the axis $w_1 = 0$. Their configuration can be described as follows: configuration of curves of type $\widetilde{\mathbb{E}}_6$ multiplied by a line and two examples of configuration of 5 surfaces described above. Each of them are glued by intersection of "pages" with a line of multiplicity 3 on $\widetilde{\mathbb{E}}_6 \times pt$.
(4) A divisor over (0,0,0).

The restriction of this family to the line $w_1 = const \neq 0$ is (up to a multiplication of a potential by a constant) an open part of Landau–Ginzburg model for the cubic threefold. Indeed,

$$(x + y + z)^3 - w_1(w_2 - t)txyz = (x + y + z)^3 - (\sqrt{w_1}w_2 - (\sqrt{w_1}t))(\sqrt{w_1}t)xyz =$$
$$(x + y + z)^3 - (w - t_1)t_1xyz,$$

where $w = \sqrt{w_1}w_2$ and $t_1 = \sqrt{w_1}t$.

The restriction to the line $w_2 = const \neq 0$ is an open part of Landau–Ginzburg model for the threefold complete intersection of a quadric and a cubic. Indeed,

$$(x + y + z)^3 - w_1(w_2 - t)txyz = (x + y + z)^3 - (w_1w_2^2)\left(1 - \frac{t}{w_2}\right)\left(\frac{t}{w_2}\right)xyz =$$
$$(x + y + z)^3 - w(1 - t_1)t_1xyz,$$

where $w = w_1w_2^2$ and $t_1 = t/w_2$.

120

On the other hand, compactified (singular) Landau–Ginzburg model for the intersection of a quadric and a cubic is

$$(t_1 + t_2)^2 (x + y + z)^3 - w t_1 t_2 xyz = t_0^2 (x + y + z)^3 - w(t_0 - t_1) t_1 xyz,$$

where $t_0 = t_1 + t_2$. In the local chart $t_0 = 1$ we get the family written down before.

Thus, after compactifying fibers of the family corresponding to $2H + H$ we get 4 additional surfaces over the w_2 axis and all together $21 = 17 + 4$ surfaces.

8.2. Hodge structures with many filtrations

We now utilize above construction of multipotential Landau–Ginzburg models from the point of view of "twistor" families. This part of the paper is highly speculative.

It is expected that Fukaya–Seidel categories with many potentials can be defined similarly to Fukaya–Seidel categories with one potential. In this case we have a divisor S of singular fibers and thimbles involved reflect not only the geometry of the fibers but the geometry of S as well. In a similar way we can associate to a Fukaya–Seidel category with many potentials a Stability Hodge Structure with a formal scheme over $M(\mathbb{P}^k, CY)$ as a fiber over zero. The following conjecture (briefly explained in Table 2) suggests a way of constructing Hodge structures with multiple filtrations.

Conjecture 8.1 (see [1]). *The mixed Hodge structure over formal scheme over $M(\mathbb{P}^k, CY)$ as fiber over zero is a mixed Hodge structure with many filtrations.*

Landau–Ginzburg moduli spaces	Nonabelian Hodge structures
$M(\mathbb{P}^1, CY)$ Landau–Ginzburg model with one potential: The fiber over zero is a formal scheme over $M(\mathbb{P}^1, CY)$, generic fibers are Stab.	Twistor family — Nonabelian Hodge Structure with one weight filtration.
Landau–Ginzburg models with k potentials	Generalized twistor families with k parameters.
The zero fiber is a formal scheme over $M(\mathbb{P}^k, CY)$, fibers (over a point in $\mathbb{C}^k$) are Stab.	Generalized multi twistor family over a k-simplex.
Extensions $M(\mathbb{P}^1, CY) \boxtimes M(\mathbb{P}^1, CY)$	Extending filtrations $u_i \boxtimes u_j$.

TABLE 2. Creating Hodge structures with multiple filtrations.

9. Birational transformations and Poisson varieties

Discussion from previous sections suggests that there is a connection between the moduli space of Landau–Ginzburg models, generators and birational geometry.

Landau–Ginzburg model	Stability
Usual Landau–Ginzburg model 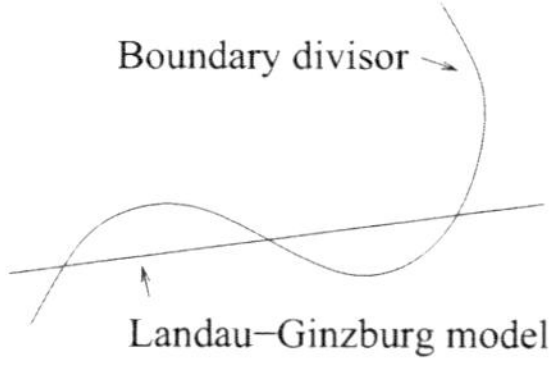 Boundary divisor Landau–Ginzburg model	Ω^3_X
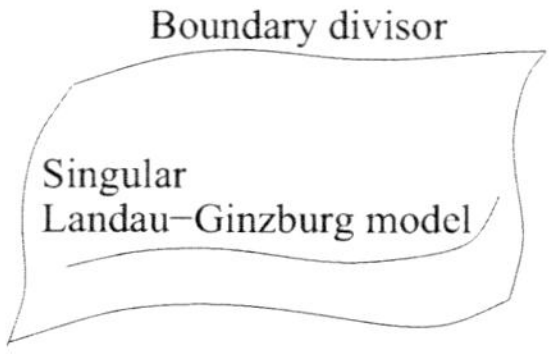 Boundary divisor Singular Landau–Ginzburg model	$\Omega^3_{X\setminus D}$ D is a divisor with stratification of singular set.
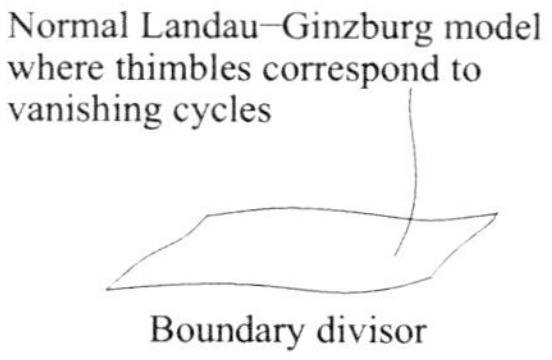 Normal Landau–Ginzburg model where thimbles correspond to vanishing cycles Boundary divisor	Ω_D $Sing\,D$ stability conditions of the vanishing cycles on D.

TABLE 3. Stability Clemens–Schmidt sequence.

Table 3 gives a version of noncommutative Clemens–Schmidt sequence for geometric stability conditions — log 3-forms. This table treats the case of three-dimensional Calabi–Yau manifolds (four dimensional Landau–Ginzburg models) but the situation in general should be rather similar. In the case at hand (three-dimensional Calabi–Yau manifold) — the stability conditions are just holomorphic 3-forms. For the quotient category (the category which produces stability conditions of the compactification) we get stability conditions to be holomorphic 3-forms vanishing in a stratified way over a divisor D. The vanishing cycles define a subcategory with its own moduli space of stability conditions and the relative (with respect to this subcategory) WCF defining an integrable system (in general a Poisson variety). The corresponding Landau–Ginzburg models can be seen as follows:

(1) The Landau–Ginzburg models associated with quotient categories are given by monotonic maps passing through an intersection of many boundary divisors in $M(\mathbb{P}^k, CY)$.

(2) The local categories of vanishing cycles are given by Landau–Ginzburg models totally within intersections of divisors.

From the perspective of generators the above splitting corresponds to splitting of the generators into the union of generators associated with the subcategory of vanishing cycles and the quotient category. In fact we get a sequence of splittings — a flag parallel to Okounkov polytopes.

These observations suggest the following conjecture, treated in [2].

Conjecture 9.1. *One-parameter families of Landau–Ginzburg models parameterize Sarkisov links.*

Recall that Sarkisov links [17] are birational maps (birational cobordisms) connecting two Mori fibrations. In our interpretation Sarkisov links become families connecting circuits. In fact we have a more general picture on the connections between moduli spaces of Landau–Ginzburg models and birational geometry. Namely we conjecture that the geometry of moduli spaces of Landau–Ginzburg models for the mirror of Fano manifold X determines its birational geometry. In particular we see a connection with relations between Sarkisov links and then relations between relations and so on. We summarize our picture in Table 4. For more details see [2], [18], [19], [20].

Sarkisov programs	Changes in the spaces of stability conditions
Commutative Sarkisov program: Sarkisov faces.	Wall crossings inside a component of stability conditions.
Non-commutative Sarkisov program: non-commutative cobordisms.	Passing from one component of stability conditions to another one.

TABLE 4. Birational geometry.

Acknowledgements. This paper came out of a talk the first author gave in Gökova, Turkey in 2011 (Sections 4, 7, 9) and discussions thereafter. We are very grateful to the organizers and in particular to S. Akbulut, D. Auroux and G. Mikhalkin for inviting us.

We thank M. Kontsevich for sharing his ideas and explaining what SHS should be. Many thanks to D. Auroux, G. Kerr, C. Diemer, D. Favero, Y. Soibelman, and T. Pantev for explaining some of the notions used in the paper. We thank S. Galkin for his explanations of cluster transformations of weak Landau–Ginzburg models. We thank N. Ilten for his explanation of embedded toric degenerations technique.

References

[1] L. Katzarkov, M. Kontsevich, T. Pantev, and Y. Soibelman, Shability Hodge structures, in preparation.

[2] C. Diemer, L. Katzarkov, and G. Kerr, Symplectic relations arising from toric degenerations, in preparation.

[3] P. Eyssedeiux, L. Katzarkov, T. Pantev, and M. Ramachandran, Linear Shafarevich Conjecture, to appear in *Annals of Mathematics*.

[4] V. Przyjalkowski, On Landau–Ginzburg models for Fano varieties, *Commun. Number Theory Phys.*, **1**(4) (2007), 713–728.

[5] V. Przyjalkowski, Weak Landau–Ginzburg models for smooth Fano threefolds, arXiv preprint, arXiv:0902.4668v2.

[6] N. Ilten, J. Lewis, and V. Przyjalkowski, Toric degenerations of Fano threefolds giving weak Landau–Ginzburg models, arXiv preprint, arXiv:1102.4664.

[7] P. Hacking, and Yu. Prokhorov, Degenerations of del Pezzo surfaces I, arXiv preprint, arXiv:0509529.

[8] N. Ilten, and R. Vollmert, Deformations of Rational T-Varieties, to appear in *Journal of Algebraic Geometry*, also arXiv preprint, arXiv:0903.1393.

[9] I. Gelfand, M. Kapranov, and A. Zelevinski, *Discriminants, resultants and multidimensional determinants*, Mathematics: Theory and Applications, Birkhauser Boston, Boston, MA, 1994.

[10] L. Lafforgue, Une compactification des champs classifiant les chtoucas de Drinfeld, *J. Amer. Math. Soc.*, **11** (1998), 1001–1036.

[11] D. Auroux, L. Katzarkov, and D. Orlov. Mirror symmetry for weighted projective planes and their noncommutative deformations, *Ann. of Math. (2)*, **167** (2008), 867–943.

[12] V. Batyrev and L. Borisov, *Dual Cones and Mirror Symmetry for Generalized Calabi–Yau Manifolds*, in Mirror Symmetry II. (eds. S.-T. Yau), pp. 65–80 (1995).

[13] K. Hori and C. Vafa, Mirror symmetry, arXiv preprint, arXiv:hep-th/0002222.

[14] V. Alexeev, Complete moduli in the presence of semiabelian group action, *Ann. of Math. (2)*, **155** (2002), 611–708.

[15] L. Katzarkov, M. Kontsevich, and T. Pantev, Hodge theoretic aspects of mirror symmetry, in From Hodge theory to integrability and TQFT: tt^*-geometry (R. Donagi and K. Wendlang, eds.), Proc. Symposia in Pure Math., vol 78, American Matematical Society, Providence, RI, 2008, 87–174.

[16] M. Ballard, D. Favero, and L. Katzarkov, The Orlov spectrum: gaps and bounds, to appear in *Invent. Math.*, also arXiv preprint, arXiv:1012.0864.

[17] V. Sarkisov, Structure of conic bundles, *Izv. RAS*, **46** (1982), num. 2.

[18] M. Ballard, D. Favero, and L. Katzarkov, Geometric Invariant Theory models and matrix factorizations, in preparation.

[19] I. Cheltsov, L. Katzarkov, and V. Przyjalkowski, Projecting Fanos in the mirror, in preparation.

[20] C. Doran, L. Katzarkov, J. Lewis, V. Przyjalkowski, Modularity of Fano threefolds, in preparation.

LUDMIL KATZARKOV, UNIVERSITY OF MIAMI AND UNIVERSITY OF VIENNA.
E-mail address: `lkatzark@math.uci.edu`

VICTOR PRZYJALKOWSKI, STEKLOV MATHEMATICAL INSTITUTE.
E-mail address: `victorprz@mi.ras.ru, victorprz@gmail.com`

List of Participants

Acu, Bahar	Middle East Technical University, Turkey
Akbulut, Selman	Michigan State University, USA
Akhmedov, Anar	University of Minnesota, USA
Akyar Moeller, Bedia	Dokuz Eylül University, Turkey
Akyol, Ayşegül	Bilkent University, Turkey
Argüz, N. Hülya	Middle East Technical University, Turkey
Atalan Ozan, Ferihe	Atılım University, Turkey
Auroux, Denis	UC Berkeley, USA
Aydın Emel	Yaşar University, Turkey
Baştürk, Saliha	Dokuz Eylül University, Turkey
Behrens, Stefan	MPIM Bonn, Germany
Bhupal, Mohan	Middle East Technical University, Turkey
Cheltsov, Ivan	University of Edinburgh, UK
Cho, Hyonjoo	University of Rochester, USA
Coşkunüzer, Barış	Koç University, Turkey
Çelik, Münevver	Middle East Technical University, Turkey
Çelik, Fatih	Koç University, Turkey
Çengel, Adalet	Middle East Technical University, Turkey
Degtyarev, Alexander	Bilkent University, Turkey
Demir, A. Sait	Istanbul Technical University, Turkey
Eden, Sinan	Boğaziçi University, Turkey
Erdal, Mehmet Akif	Bilkent University, Turkey
Erdoğan Demir, Sultan	Bilkent University, Turkey
Etgü, Tolga	Koç University, Turkey
Finashin, Sergey	Middle East Technical University, Turkey
Gayet, Damien	Institut Camille Jordan, France
Genç, Özhan	Middle East Technical University, Turkey
Georgieva, Penka	Stanford University, USA

Greene, Joshua Evan	Boston College, USA
Güğümcü, Neslihan	Koç University, Turkey
Gürbüzer, Kaan	Dokuz Eylül University, Turkey
İşal, Hanife	Middle East Technical University, Turkey
Itenberg, Ilia	IRMA, Université de Strasbourg, France
Johns, Joe	Max-Planck Institute, Germany
Kalafat, Mustafa	Middle East Technical University, Turkey
Katzarkov, Ludmil	University of Miami, USA
Kişisel, Özgür	METU North Cyprus Campus, North Cyprus
Koçak, Şahin	Anadolu University, Turkey
Lebedeva, Nina	Steklov Institute of Mathematics, Russia
Leung, Conan	Chinese Univ. of Hong Kong, Hong Kong
Mikhalkin, Grigory	Université de Genève, Switzerland
Ozan, Yıldıray	Middle East Technical University, Turkey
Önder, Turgut	Middle East Technical University, Turkey
Özbağcı, Burak	Koç University, Turkey
Özer, Arda Buğra	Atılım University, Turkey
Pamuk, Mehmetcik	Middle East Technical University, Turkey
Pamuk, Semra	Middle East Technical University, Turkey
Petrunin, Anton	Pennsylvania State University, USA
Przyjalkowski, Victor	Steklov Mathematical Institute, Russia
Sağlam, K. Nur	Middle East Technical University, Turkey
Salur, Sema	University of Rochester, USA
Sarıoğlu, Celal Cem	Dokuz Eylül University, Turkey
Shapiro, Michael	Michigan State University, USA
Stanislav, Smirnov	Université de Genève, Switzerland
Şahin, Mehmet	Istanbul, Turkey
Taştan, Hakan Mete	Istanbul University, Turkey
Tange, Motoo	RIMS, Kyoto University, Japan
Todd, Albert	University of Rochester, USA
Ünal, İbrahim	METU North Cyprus Campus, North Cyprus
Yan, Min	Hong Kong University of Sci. and Tech., Hong Kong
Yasui, Kouichi	Hiroshima University, Japan
Zhang, Weiyi	University of Michigan, USA

27815526R00077

Made in the USA
Columbia, SC
30 September 2018